DISCARD

NUCLEAR STRATEGY AND STRATEGIC PLANNING

355.033573 Gray
G778n   Nuclear strategy
          and strategic planning

Glendale College Library

NUCLEAR STRATEGY AND STRATEGIC PLANNING

Colin S. Gray

Philadelphia Policy Papers

FOREIGN POLICY RESEARCH INSTITUTE
Philadelphia, Pennsylvania
1984

Copyright © 1984 by the Foreign Policy Research Institute.
All rights reserved. Manufactured in the United States of America.

355.033573
G778n

Library of Congress Cataloging in Publication Data
Gray, Colin S.
  Nuclear strategy and strategic planning.

  (Philadelphia policy papers)
  Includes bibliographical references.
  1. United States—Military policy—Congresses.
  2. Atomic warfare—Congresses. 3. Strategy—Congresses.
  4. Military planning—United States—Congresses.
  I. Title. II. Series.
  UA23.G779 1984   55'.0335'73   83-20800
  ISBN 0-910191-07-7

Foreign Policy Research Institute
3508 Market Street, Suite 350
Philadelphia, PA 19104

The Foreign Policy Research Institute is a publicly supported, nonprofit, tax-exempt corporation as described in Section 170 (b) (1) (A) (vi) of the Internal Revenue Code. All contributions to the Institute are tax deductible.

FOREIGN POLICY RESEARCH INSTITUTE

The Foreign Policy Research Institute is a nonprofit organization devoted to scholarly research and analysis of international developments affecting the national security interests of the United States. The Institute's major activities include a highly diverse publications program; seminars, workshops, and conferences for business, government, and academic leaders; and important research projects for a wide range of government agencies.

The Institute's research and publications programs are designed to:
- Identify and explore political, military, and economic trends in the international system;
- Analyze the fundamental issues facing U.S. foreign policy;
- Suggest guidelines for U.S. foreign policy that contribute both to American security and to the development of a stable international order.

To accomplish these purposes, the Institute:
- Maintains a staff of specialists in the field of international relations as well as an extensive library of pertinent periodicals, books, and information files. The Institute's library is open to use by scholars, students, and the general public;
- Draws upon the expertise of leading academicians and persons in public life who are concerned with international affairs;
- Convenes and publishes the results of conferences and symposia of U.S. and foreign experts on contemporary international problems.

The products of the Institute's research program include the following publications:
- **ORBIS:** a quarterly journal of world affairs, widely recognized as a leading forum for research in the field of international and strategic studies;
- **Foreign Policy Issues:** a book series (in cooperation with Praeger Publishers) of important studies written by authorities in the field of international and strategic studies;
- **Philadelphia Policy Papers:** analyses of timely issues and developments having serious implications for U.S. foreign and defense policy;
- **Research Reports:** sponsored by various institutions, including agencies of the U.S. government.

The opinions expressed in publications of the Foreign Policy Research Institute are those of the authors and should not be construed as representing those of the Institute.

OTHER TITLES IN THE PHILADELPHIA POLICY PAPER SERIES

The War Powers Resolution: Its Implementation in Theory and Practice. By Robert F. Turner, with a foreword by Senator John G. Tower (1983) $4.95.

"an excellent . . . study" *The Wall Street Journal*

Western Europe's Middle East Diplomacy and the United States. By Adam M. Garfinkle (1983) $4.95.

"lively and provocative" *Middle East Journal*

Economic Sanctions in U.S. Foreign Policy. By Shaheen Ayubi, Richard E. Bissell et al. (1982) $3.95.

Ground Rules: Soviet and American Involvement in Regional Conflicts. By Joanne Gowa & Nils H. Wessell (1982) $3.95.

"long overdue" *International Affairs* (London)

FORTHCOMING

The Politics of the Nuclear Freeze. By Adam M. Garfinkle (1984) $5.95.

To Valerie

CONTENTS

Foreword *William R. Van Cleave*   xiii
Preface   xvii

1. The Ambivalence of Doctrine   1
2. Policy Guidance, War Plans, and Strategy   17
3. Nuclear Weapon Policy: Evolution and Debate   37
4. Nuclear Strategy: The Range of Choice   55
   - Option One: Mutual Assured Vulnerability   59
   - Option Two: Mutual Assured Vulnerability with Targeting Flexibility   67
   - Option Three: Counterforce and Countercontrol Preeminence with Recovery Denial   71
   - Option Four: Damage Limitation for Deterrence and Coercion   79
   - Option Five: Damage Limitation with Defense Dominance   86
5. Nuclear Strategy: A New Consensus?   109

FOREWORD

Over the past decade or so, no one has written more knowledgeably, lucidly, or cogently on matters of national strategy, nuclear policy, or arms control than Colin Gray. When we first began to hear from Dr. Gray, his thoughts on strategy were incisive but seemed iconoclastic, given the prevailing academic and official views. Since then, many of his views have become more widely accepted, and certainly more compatible with official U.S. nuclear policy. Of course, as Dr. Gray points out in this monograph, that policy has evolved considerably over the past fifteen years.

This work is an attempt to provide better understanding of the issues involved in the current debate over the doctrine, concepts, and objectives for U.S. strategic nuclear forces. As such, it is a timely and useful contribution to the literature on nuclear strategy and arms control for several reasons.

First, it provides an even-handed summary and elucidation of the major elements and positions in the contemporary debate over U.S. nuclear strategy. It is more than mere exegesis and exposition, however; it provides a logical framework for examining alternatives. The work is definitely not a polemic. This is not to say that Dr. Gray is reticent about expressing his own views or presenting his own preferred strategy. Rather, it is that he goes out of his way to treat alternative or opposing points of view fairly, even generously. In fact, he seems to reject none of the various positions or schools of thought totally, but finds what he believes to be valid points in all of them. Even "the more radical critiques of extant U.S. nuclear strategy," he says, "do identify some genuine problems." Moreover, "one purpose of this study is to show the fragility of all arguments with respect to choices among nuclear strategies." Consequently, students of the subject holding differing views should find this work acceptable and valuable.

Second, and at least equally useful to students—particularly, I think, graduate students—the work is eminently scholarly. As regards the literature on nuclear war and strategy, Dr. Gray is clearly a bibliophile with encyclopedic knowledge. His copious endnotes provide an excellent source of suggestions for further reading, and I would expect that many professors will enrich their course reading lists by cribbing freely from them. I certainly intend to do so.

Third, Dr. Gray presents and examines five options for nuclear strategic policy. These options, ranging from mutual assured vulnerability to a defense-dominant, damage-limiting posture, are progressively more ambitious in their objectives and more demanding in their force requirements. While they could be presented as purely logical constructs, each, in fact, has adherents in the open literature. The categorization is useful and not overly simplified; the options overlap, but each is sufficiently distinct to merit separate treatment. None is presented as a "straw man," and each is explained and examined in detail. Dr. Gray's technique is to break each position down into a set of key tenets, beliefs, or presumptions and then to submit each to critical commentary. He uses this approach in discussing his own preferred strategy as well.

This treatment, contained in chapter 4, is the heart of the work. The first three chapters provide a backdrop for chapter 4, and the final chapter is less a conclusion than a compendium of thoughts suggested, particularly, by the Scowcroft Commission Report, which Dr. Gray finds far more congenial and compelling than I.

Finally, Dr. Gray's argument for his own strategic preference is a major contribution. Dr. Gray correctly identifies current official U.S. *declaratory* policy as his Option Three (counterforce preeminence with recovery denial). He regards this as proper, and as a decided evolutionary improvement, but prefers that it be a transition to his own preference, which is to add a healthy damage-limiting capability through U.S. homeland defenses. His argument strikes me as sensible and practical in both a technical and a political sense. Dr. Gray points out that, despite the diversity of options and views he presents, there is also a fair degree of consensus on the basic elements of current U.S. nuclear strategy and its objectives. Despite the radical views of those whom Albert Wohlstetter long ago termed "spokesmen of the apocalypse," the mainstream of strategic thought in the United States today endorses a policy of deterrence with strong elements of counterforce, control, and damage limitation. There is reason to believe that public opinion also supports such a policy, despite the best efforts of the media (e.g., the ABC drama "The Day After") to instill the belief that nuclear devastation cannot be prevented in the event of any type of nuclear war and that attempts to control or limit it only promote such a war.

In my own view, Dr. Gray accepts too much of the argument that the use of strategic nuclear weapons cannot be controlled (true, he says "controlled closely," and he uses "improbable" rather than "cannot"). I am unsure whether the net effect of this concession is to bolster his argument for improved defenses or to weaken his case for damage limitation, but I would not make the concession in either case.

Contemporary critics of planning for limited and controlled strategic nuclear operations have advanced beyond the old article-of-faith assertion of inevitable escalation once the nuclear firebreak has been breached. They now make technical arguments that control over the use of U.S. nuclear forces would be lost, because, for example, of a breakdown in communications. That U.S. C^3 (command, control, and communications) is far too fragile, and probably unenduring, is clear. But that its disruption would tend toward unlimited, rather than limited, use of strategic nuclear weapons is far from clear.

In effect, what this argument suggests is that *if* the Soviet attack is large-scale to begin with, and designed to destroy as much U.S. strategic power, command authority, and communications as possible, then plans for tightly controlled limited responses fail or become irrelevant. Largely, that is tautological. Of course, the outcome of such a successful strike—and its purpose in the first place—would be to deny, to deter, or to disrupt and degrade any U.S. retaliation. And to the degree that it were successful, any U.S. retaliatory attack would be limited.

Of course, it is precisely a Soviet capability to do this that we must avoid. Critics of limited war options pose the prospect of such a successful attack as an argument against the development of plans and capabilities for controlled retaliatory operations.

Yet, if these critics had their way, they would in fact make it *easier* for the Soviets to accomplish their aims in such an attack. For deterrence, it is imperative that we erase, not perpetuate, those vulnerabilities and deny the Soviets the capability for such a disruptive counterforce attack. It is also imperative—and Dr. Gray agrees—that we have the capability for controlled responses suitable to the situation at hand. If control over the use of nuclear weapons in a way that might meaningfully limit damage and bolster deterrence is "improbable" because of technical weaknesses, the obvious answer is to correct the weaknesses. One of the principal reasons for determining strategic objectives is to provide guidance for force modernization and improvement—to identify what needs to be done. To accept that "it is improbable that strategic-nuclear use could be controlled closely and in a purposive manner" strikes me as discouraging preparation. That is clearly not Dr. Gray's intention.

My final comments have to do with the gap between official strategic doctrine and actual force capabilities, to which I have just referred. U.S. declaratory strategic doctrine is, indeed, Dr. Gray's third option, but U.S. force capabilities, at best, coincide with his more modest Option Two (mutual assured vulnerability with targeting flexibility). Current strategic force programs, at least through the 1980s, will not change that situation. In my own view, the gap between our doctrine and objectives, on the one hand, and our capabilities and programs, on the other, is a very serious problem. What do we do with a surviving force, given Soviet attack capabilities, that would in all probability be much closer to McNamara's canonical 400-equivalent-megaton "assured destruction" capability than to the force required to meet present targeting objectives? What does having such a force mean for deterrence and for strategic or crisis stability, or for extended deterrence and foreign policy? While having a strategic plan and doctrine beyond current capabilities is useful for force planning purposes, what is its relevance to the actual capabilities and realistic options that would exist should deterrence fail at the strategic nuclear level? Would we then be forced, willy-nilly, to abandon our declared strategic doctrine? Or are there ways—less than optimal, or even adequate, to be sure—to accomplish a useful portion of the current plans and objectives even with relatively inferior forces?

This valuable study by Dr. Gray provides essential background for those interested in finding answers to crucial questions such as these, which are shaping the contemporary debate on nuclear strategy.

William R. Van Cleave
University of Southern California

PREFACE

We are nearing the end of the fourth decade of the nuclear age, and profound and even bitter dispute persists over fundamental aspects of nuclear strategy and planning. There can be no true experts on a phenomenon no one has experienced firsthand—bilateral nuclear war. In the field of nuclear strategy there is no operational history for scholars to study or for official practitioners to draw lessons from. All our knowledge in this area is derived from the study of hypothetical events. The merit of an idea concerning nuclear strategy can be tested only in terms of another idea.

It might be supposed that even though we are ignorant of the realities of nuclear war we should be knowledgeable concerning the foundations and maintenance of nuclear peace. Has not nuclear deterrence been a spectacular success for nearly forty years? Notwithstanding the unknown depth of our ignorance about nuclear war, it is far more important that we understand how to keep the peace. In that regard it would seem that the nuclear powers must have been doing much that was right.

Although history tells us that the conditions for nuclear peace have persisted since the late 1940s, it is less helpful as a guide to *what deters*. The nuclear deterrence system may be said to have worked for more than three decades, but the system may not truly have been tested. One cannot demonstrate that war would have occurred but for the presence of nuclear weapons, or that war would have occurred but for a particular state of the strategic balance.

The most dangerous crises of the nuclear age—the protracted confrontation over Berlin (1958-61) and the Cuban Missile Crisis of October 1962—were at least as much products of the nuclear balance (and perceptions of imbalance) as they were beneficiaries of the urge to be cautious induced by fear of nuclear war. Over Berlin, Nikita Khrushchev sought political gain through nuclear intimidation; over Cuba it would seem in retrospect that a deficiency in strategic nuclear assets was the spur to Soviet forward nuclear deployment.

Even though we cannot with authority answer the question

of what would deter a Soviet leadership (currently unknown) at a particular time (again unknown) with reference to a historically unique issue or set of issues (unknown yet again), U.S. and NATO defense planners are obliged to provide an answer and to build forces that could give adequate expression to the threats required by the strategy so identified. Critics of nuclear strategy and of nuclear strategists like to point to the difficulties of the field of strategic studies and to the extreme possibilities that might well render the labors of the nuclear defense planners sheer vanity or wishful thinking. For example, Fred Kaplan concludes his book-length study of nuclear strategists with the following assessment:

> The precise calculations and the cool, comfortable vocabulary were coming all too commonly to be grasped not merely as tools of desperation but as genuine reflections of the nature of nuclear war.
>   It was a compelling illusion. Even many of those who recognized its pretense and inadequacy willingly fell under its spell. They continued to play the game because there was no other. They performed their calculations and spoke in their strange and esoteric tongues because to do otherwise would be to recognize, all too clearly and constantly, the ghastliness of their contemplations. They contrived their options because without them the bomb would appear too starkly as the thing that they had tried to prevent it from being but that it ultimately would become if it ever were used—a device of sheer mayhem, a weapon whose cataclysmic powers no one really had the faintest idea of how to control. The nuclear strategists had come to impose order—but in the end, chaos still prevailed.[1]

There is a great deal of intellectual order in nuclear debate and nuclear planning, and to the knowledge of this author no one disputes the potential for chaos or even "sheer mayhem" should nuclear weapons ever be used. Critics such as Kaplan appear to believe they have said something profound and helpful when they remind us of the all-too-obvious possibility of nuclear catastrophe that lurks in the present security system. Understandably, responsible officials tend to be impatient with those who tell them what they know already—that they have a problem. The officials cannot lament the difficulty of relating the energy yield of nuclear weapons to the securing of political objectives—that is, the difficulty of accommodating nuclear weapons in *strategy*—and throw their hands up in despair. Unlike the extragovernmental critic, they must do the best they can in making security sense of a weapon that cannot be disinvented.

Chaos and mayhem may attend and follow actual nuclear use, but governments are obliged to design strategy and tactics for the threat and employment of these weapons. So long as politicians and officials do not forget the potential for chaos and mayhem that exists in the nuclear arsenal—a highly improbable error—there is everything to be said in favor of planning for controlled and limited employment of nuclear weapons in support of foreign policy objectives. If there has to be planning for nuclear defense, as there must, would it be prudent to plan only for nuclear Armageddon? This question virtually answers itself.

This study is designed to shed light on two closely related aspects of nuclear-weapon policy: (1) the realities of strategic planning (including the ambivalence of doctrine) with particular reference to plans for employment of nuclear weapons, and (2) the range of choice over nuclear strategy which is available theoretically. Designers of nuclear strategies are frequently accused of indifference to, or ignorance of, real-world constraints—what Karl von Clausewitz identified as the factor of "friction."[2] The first two chapters of this study acknowledge and describe the "real world," where policy guidance may be ambivalent and where war plans lose their value after the first clash of arms.

The "real world" mandates that policymakers and advisers be as clear and purposive as possible in the ideas they discuss and promote. There are some different core ideas in the field of nuclear strategy, and the distinctions among them have major implications for foreign policy, force structure, and the risk of war. In chapter 4 the author's preference for nuclear strategy Option Four (damage limitation for deterrence and coercion) is plain, but the purpose of this study is not to seek converts for a particular vision of the requirements for U.S. nuclear-weapon policy. Each of the five strategy options discussed in chapter 4 is presented in a pro/con format that—when read in light of the more discursive treatment of "real world" difficulties presented in the preceding chapters—should provide a basis for objective assessment. The final chapter provides an assessment of the contemporary nuclear policy debate, with a view to determining just where and why rival policy advocates agree and disagree.

Colin S. Gray
Fairfax, Virginia
July 1983

NOTES

1. Fred Kaplan, *The Wizards of Armageddon* (New York: Simon & Schuster, 1983), pp. 390-91.
2. Karl von Clausewitz, *On War,* ed. and trans. Michael Howard and Peter Paret (Princeton: Princeton University Press, 1976), p. 119.

Chapter 1

THE AMBIVALENCE OF DOCTRINE

The first, the supreme, the most far-reaching act of judgment that the statesman and commander have to make is to establish . . . the kind of war on which they are embarking; neither mistaking it for, nor trying to turn it into, something that is alien to its nature. This is the first of all strategic questions and the most comprehensive.

—Karl von Clausewitz

If, as Clausewitz so justly said, war is a continuation of national policy, so also are war plans.

—Barbara Tuchman

A weapon system can be assessed for its strategic merits only in terms of criteria provided by strategy, while the appropriateness of strategy can be assessed only in terms of national security policy or grand strategy. The unique strategic characteristics of a particular weapon are of little interest if those characteristics are undesired for deterrent effect or for actual performance in combat.

The U.S. government has reportedly decided that, in order to have the needed deterrent effect on Soviet minds, the United States must have the capability to conduct protracted conflict (including protracted nuclear conflict).[1] At the theoretical level, few people are confused about the relationship between deterrence and so-called war-fighting capability.[2] In the public debate over nuclear strategy today, there is disagreement over the requirements of deterrence, which often leads rival commentators to debate weapons when they should be debating strategic theory. Alternative theories of deterrence are explored in chapter 4. At this point the principal schools of thought will be identified, plausible trends in nuclear strategy design will be specified, and the range of strategy choice will be summarized. The difficulty the U.S. government has explaining its deterrence story, providing rationales for individual weapon and weapon-support programs, and providing clear policy guidance for war planning appears to stem in part from some ambivalence over matters of fundamental strategic doctrine.

First it is necessary to identify the three broad schools of deterrence thinking that dominate the policy debate:

1. *Societal Punishment.* In the several variants of this view, the Soviet Union is deterred by the prospect of suffering societal punishment on a massive scale. Defense Department analysts in the mid 1960s identified the knee of the marginal cost curve of effort expended to damage inflicted at around 400 one-megaton equivalents.[3] If it is a given that for the Soviets "victory," or even the avoidance of defeat, requires the survival of a viable Soviet society, the promise of a 400 equivalent megaton laydown by the United States is enough to deter. The difference between massive damage and more massive damage is not of interest from the perspective of deterrence or of likely physical and social effects.

2. *The "Countervailing Strategy."* In this view, the Soviet Union is deterred when confronted by threats that both focus on the highest of official values and have some credibility about them with respect to likelihood of translation into action. The United States deters by threatening to thwart Soviet strategy—deny victory to the Soviet Union—by striking at the war-waging structure and instruments of the Soviet state. Credibility is lent to the prospect of execution by targeting plans that are designed to preclude, or minimize, unwanted collateral damage.[4]

3. *The "Prevailing Strategy."* In this view, the Soviet Union is deterred when the United States is able both to impose military—and hence political—defeat on Soviet arms and to secure the achievement of Western political purposes at a military, economic, and social cost commensurate with the stakes of the conflict.[5] Because Soviet style in the waging of war is beyond our control, though perhaps not our influence, the limiting of damage to Western societies cannot be allowed to repose substantially in hopes for reciprocation in targeting restraint.[6] A United States that believes it must in extreme circumstances be able to "prevail" if it is to deter must set its planning sights considerably beyond developing a defense posture that will simply deny victory to the enemy. To prevail in stressful circumstances the United States must be able to defend itself against nuclear attack.[7] In this per-

spective, credibility for deterrent effect flows not so much from flexibility in strategic planning or from the reciprocated restraint it is hoped would be encouraged by the extreme care with which targets would be struck.[8] Instead, credibility would flow from Soviet belief, or strong suspicion, that the United States could fight and win the military conflict and hold down its societal damage to a tolerable level.

All three deterrence schools come in stronger or weaker versions, and each school accommodates the need for and feasibility of flexibility in strategic targeting (even assured destruction can be effected in installments).[9] Also, each school is compatible with strategic planning requirements that have a wide range of endurance goals for forces under wartime conditions.[10]

The Reagan administration is in reality endorsing strategic programs for the "countervailing strategy," but it often speaks as if it either is endorsing "prevailing strategy" *as policy,* or at least as if it favors it theoretically.

The three schools of nuclear-policy thinking identified here comprise the scope of the political debate today, but their existence tends to obscure some important distinctions and refinements. The discussion in chapter 4 of the range of choice in nuclear strategy is concerned more with the defense-analytical integrity of arguments than with the general rival clustering of positions. Options One and Two in chapter 4 are compatible with school one, "societal punishment"; Option Three is synonymous with school two, the "countervailing strategy"; Option Four matches school three, the "prevailing strategy"; and Option Five ("damage limitation with defense dominance") is too distant a technical prospect to be a very active player in the U.S. strategy debate as yet.

This study endorses the view that punishing Soviet society is unnecessary for deterrence, suicidal in its logical consequences, incredible as a threat (or sequence of threats), and unethical. Nonetheless, because of the geographical proximity of Soviet war-waging assets to urban-industrial areas,[11] Desmond Ball is probably correct in asserting that even though the October 1980 nuclear weapons employment policy abandons the requirement that the U.S. strategic nuclear forces have as their priority mission

70 per cent destruction of Soviet economic recovery and war-supporting assets, "it still remains the case that the top 300 urban-industrial areas in the Soviet Union will continue to receive about the same amount of nuclear fire power and to suffer about the same amount of damage."[12]

Whatever the collateral damage to civilian assets resulting from a nuclear campaign, the Reagan administration does not endorse the following description of the first priority of U.S. nuclear targeting strategy offered by Harold Brown early in 1980:

> We need, first of all, a survivable and enduring retaliatory capability to devastate the industry and cities of the Soviet Union . . . . What has come to be known as assured destruction is the bedrock of nuclear deterrence, and we will retain such a capability in the future. It is not, however, sufficient in itself as a strategic doctrine.[13]

The distinction between military (counterforce) and civilian (countervalue) targets would in many cases be more theoretical than real and certainly could not be appreciated by a recipient nation whose attack-assessment assets had been degraded or destroyed. Alleged distinctions between "war-fighting" and "deterrent" strategies are wrong on at least three major counts. First, they reflect a false opposition. The true distinction is between a theory of deterrence through defense (the threat to repel and/or defeat the enemy) and deterrence through the threat of punishment. Second, one can wage a war by attacking enemy forces and/or by attacking the society that sustains and provides those forces—all strategies for employment of military power are "war-fighting" strategies. Third, nuclear warheads ranging from 0.1 to 1.0 megaton, delivered over 6,000-to-8,000-mile trajectories, are not surgical instruments. U.S. nuclear strategy may not require that Soviet society be threatened or attacked, but there is no way for a major counterforce campaign to avoid the collateral punishment of Soviet society.

Some administration spokesmen attracted criticism when they defined the requirements of U.S. nuclear strategy.[14] It should be understood that although the United States can, and much prefers to, prevail without war (that is, to deter successfully), and could prevail in that U.S. political purposes might be secured at the cost of relatively little damage (*for a nuclear war*) to U.S. (and allied) society, a theory of deterrence through the ability to "prevail" should rest on a policy story that is more substantial than hopes for reciprocated targeting restraint (i.e., intrawar deterrence). To talk "prevailing" (meaning that they lose *and we win*), but practice "countervailing" (meaning that they lose and we also may lose), invites confusion in public debate and in strategic planning.[15]

Four principal sources of the contemporary uncertainty appear to have promoted confusion in debate. First, there is difficulty distinguishing between the necessary and the desirable. Although the United States and the North Atlantic Treaty Organization (NATO) have judged it essential that the Soviets be denied victory, they have not agreed on whether it is necessary to impose defeat. Some scholars suggest that denial of victory would translate into a political defeat that would imperil the stability of the Soviet imperium.[16]

Second, so-called war-fighting theories of deterrence may be held to indicate a wide range of possible defense requirements. There is a significant difference between a U.S. strategy that merely focuses on military objectives, to ensure that a war is waged in the traditional manner (to ignore strategic bombing in World War II and the U-boat campaigns of both world wars) of force on force, and a U.S. strategy that plans seriously for defeat of the enemy. The distinction may be illustrated with reference to NATO ground and tactical air forces. Those forces guarantee a major war in the event of a Warsaw Pact invasion, but they are not intended by their own effort to defeat an invasion.[17] They increase the risk of a much wider and more destructive war. There is a difference between deploying strategic nuclear forces and C^3I (command, control, communications, and intelligence) support systems that simply will give a good account of themselves, and deploying strategic nuclear forces that offer the prospect of defeating the enemy.

Third, while common sense suggests that Soviet leaders should be deterred by the prospect of losing a war, less clear thinking manifests itself with respect to the possibility of the United States' *winning* a war. If Soviet leaders believe that the United States could achieve its political goals at tolerable cost, then they will not perceive the United States as being self-deterred by visions of the end of the world from pressing its cause to the point of military decision.

Fourth, an appearance of intellectual and policy confusion and uncertainty is produced if an administration believes that nuclear war is survivable, and even winnable in a politically meaningful sense, but knows or suspects that the policy story and the associated dollar costs will not be accepted in the United States or abroad. Every administration has had a variant of this problem and has chosen not to explain the full ramifications of its strategic thinking. The facts are that since the late 1940s the U.S. government has made provision for waging nuclear war and that there is no nuclear strategy of any kind that would not produce casualty levels abroad and at home that would exceed by far the historical norm in the American experience.[18] Rival schools of thought on nuclear strategy all agree that deterrence is their aim, but they differ over the kind of military preparation necessary to make nuclear threats credible and over operational strategy if deterrence should fail.

Clausewitz stated the problem more clearly than contemporary strategists: "The most far-reaching act of judgment that the statesman and commander have to make is to establish . . . the kind of war on which they are embarking." The other quotation with which this chapter opened, from Barbara Tuchman, similarly points to a fundamental issue: "If . . . war is a continuation of national policy, so also are war plans." The U.S. government today seems to be victim of a tension that all commentators on and theorists about nuclear strategy feel but nonofficials can ignore. On the one hand, there is a feeling (rather than a judgment) that nuclear war would be so different that "its nature" (in Clausewitz's term) cannot really be contained and harnessed by military plans.[19] This sense of awe in contemplation of the unknown[20] is married to a sense of unreality that is not unprecedented in modern history. The century that saw no general European war between 1815 and 1914 produced a "peacetime mentality" that required a succession of acute political crises and several months of actual combat to break down.[21] On the other hand, U.S. defense officials must daily penetrate the unknown in planning for the conduct of nuclear operations and judging and justifying new military programs that executing those plans requires.

In strategic logic it is not difficult to plan a nuclear campaign so that the execution of military operations serves the goals of high policy. But in political logic it may well be exceedingly difficult to conduct a nuclear war in such a way that political purposes can be served at tolerable cost. The U.S. defense planner knows this but is obliged to ignore it. The problem is with both objective reality and subjective reality. Objective reality is that nuclear weapons exist, that plans must be designed for their employment, that policy must guide those plans, and that the whole edifice should appear credible in order to show ourselves and others that the U.S. nuclear deterrent is no mere bluff.

But how serious are we about planning for nuclear war? How serious should we be?[22] The Reagan administration, reflecting a genuine ambivalence among defense professionals of both parties, appears to be hovering between endorsement of the countervailing strategy that it inherited and a prevailing strategy that has greater intellectual and policy appeal but has associated political and economic costs that are formidably dissuasive.

The public has not been told—and perhaps the defense community itself has not fully realized—that there are two distinct kinds of justification for development of what is called a warfighting capability. First, there is widespread agreement that a perceived ability to wage a nuclear counterforce (including counter C^3) war is probably critically important for the credibility of nuclear threats and hence for the stability of prewar deterrence.[23] Second, a vigorous war-fighting capability may be justified on the ground that the nuclear deterrence system linking East and West is not foolproof or accident-proof.[24] Although nuclear-armed states have to date behaved circumspectly in their mutual relations, to a degree that is unprecedented in the recorded history of international politics, it is possible that a condition could arise either where one party would be beyond deterrence or where a succession of crises produced a conflict that neither party had the skill or determination to prevent.[25]

While maintaining that the United States should not develop and sustain a defense posture that makes war more probable (a caveat that raises a host of important policy-relevant issues that transcend the bounds of this inquiry),[26] there is a rationale for nuclear war-fighting capability that is distinct from deterrent reasoning. Nuclear war is possible, and the U.S. government owes it to generations of Americans—past, present, and future—to make prudent defense preparations to limit damage to domestic American values to the extent feasible in the event of nuclear war.

The ambivalence of current strategic doctrine and policy is best demonstrated by reference to the issue of whether a nuclear war could be won. In his 1982 letter to thirty domestic and forty foreign newspaper editors, the secretary of defense wrote, "We do not believe there could be any 'winners' in a nuclear war."[27] This apparently clear statement can be interpreted in two ways. The first interpretation is that both sides would lose, though there may be degrees of losing and those degrees could be important. For example, even if the United States were to suffer in the low range of 5 to 15 million fatalities in a nuclear war that was terminated with the dissolution of the Soviet state and U.S. sponsorship of a (or several) successor regime(s), Americans might judge the human cost to have been so high that it mocked the claim of political-military victory.[28] In the second interpretation, the statement could mean that both sides would lose in that neither would achieve its political purposes *at tolerable cost*. Does this mean that even though U.S. nuclear employment would thwart Soviet strategy it could not achieve the political purposes intended?

While in response to signs of public political alarm officials are talking down the possibility of winning a nuclear war, do they mean to imply that nuclear weapons have "utility [only] in nonuse," as Bernard Brodie contended?[29] This is not an academic point. There is practical significance attached to resolution of the issue of whether the U.S. government believes that some kinds of nuclear use may, on rare occasions, serve U.S. political ends. For more than thirty years a heavy U.S. focus on the needs of prewar or crisis time *deterrence* has given short shrift to considerations bearing on the actual conduct of nuclear war.[30] The range of inimical actions that the United States might reasonably aspire to deter by nuclear threat has undoubtedly narrowed as a result of the dramatic shift in the strategic balance since the mid 1960s.

But the United States and its allies do, and must, ask of nuclear weapons considerably more than that they function simply as a counterdeterrent, inhibiting Soviet nuclear initiatives. As a matter of geostrategic necessity and elementary prudence, U.S. policymakers envisage the possibility of actual nuclear employment for the purpose of restoring deterrence. This author agrees with the Scowcroft Commission (the President's Commission on Strategic Forces) that "deterrence . . . requires military effectiveness,"[31] but whatever one's theory of the requirements of deterrence, it is difficult even to conceive of a military capability functioning as a deterrent if it is bereft of a theory of controlled execution in support of political purpose.

There have always been ambivalence, lack of clarity, and confusion in U.S. and NATO-allied thinking on nuclear strategy. But frustrating though it can be to lower-level officials who must implement policy, such thinking may have positive value for the stability of deterrence. For example, the U.S. government may not be certain that it could distinguish a large counterforce attack from a countervalue attack, and it may be very uncertain about how it would, or even whether it could, respond in the event of such uncertainty. Soviet anticipation of such U.S. uncertainties should promote caution in Moscow.[32] However, although uncertainty is valuable for stability in deterrence, and while one may make a strategic virtue of the inevitability of uncertainty, this good idea easily becomes much less good if taken too far. While it would be desirable for the Soviet leadership to be uncertain about the character of probable American behavior, the U.S. government should itself know what it is about. As a deterrent virtue, uncertainty can provide a too convenient cover for doctrinal confusion and muddled thinking.

Disciplined discussion of the value in particular nuclear strategies, as provided in chapter 4, must be balanced by and founded on appreciation of some critically important general truths that pertain both to strategy and strategic planning, and to strategy and strategic planning in the unique historical context of the United States in the 1980s. To this subject we now turn.

NOTES

1. See Richard Halloran, "Pentagon Draws up First Strategy for Fighting a Long Nuclear War," *New York Times,* May 30, 1982, pp. 1, 12; and Michael Getler, "Weinberger Speech: Pentagon Acts to Clarify Position on Nuclear War," *Washington Post,* June 4, 1982, p. 4. The unfavorable publicity in 1982 that surrounded the idea of protracted nuclear war led the Department of Defense in 1983 to be circumspect virtually to the point of silence in its public discussion of the subject. See the discreet paragraph in Caspar W. Weinberger, Secretary of Defense, *Annual Report to the Congress, Fiscal Year 1984,* February 1, 1983, p. 35; see, also, the reference on p. 51. Official Washington seems to have decided that "enduring deterrence" is politically more palatable language than "protracted war."

2. The classic theoretical exposition is Gleen Synder, *Deterrence and Defense: Toward a Theory of National Security* (Princeton: Princeton University Press, 1961). For a recent treatment, see Colin S. Gray, *War-fighting for Deterrence* (Fairfax, Va.: National Institute for Public Policy, July 1983).

3. See Alain C. Enthoven and K. Wayne Smith, *How Much Is Enough? Shaping the Defense Program, 1961-1969* (New York: Harper & Row, 1971), pp. 207-8; and Jerome H. Kahan, *Security in the Nuclear Age: Developing U.S. Strategic Arms Policy* (Washington: Brookings Institution, 1975), pp. 200-23.

4. The countervailing strategy, which expresses the mainstream opinion of the defense-professional community, is well described in Harold Brown, *Department of Defense Annual Report, Fiscal Year 1981,* January 29, 1980, pp. 65-70; Harold Brown, Remarks prepared for delivery at the Naval War College, Newport, Rhode Island, August 20, 1980; U.S. Congress, Senate, Committee on Foreign Relations, *Nuclear War Strategy, Hearing,* 96th Cong., 2d sess., 1980 (sanitized and published on February 18, 1981); Harold Brown, *Department of Defense Annual Report, Fiscal Year 1982,* January 1981, pp. 38-45; and Walter Slocombe, "The Countervailing Strategy," *International Security,* vol. 5, no. 4, Spring 1981, pp. 18-27.

5. See Colin S. Gray and Keith B. Payne, "Victory Is Possible," *Foreign Policy,* no. 39, Summer 1980, pp. 14-27. For a critique of this idea, see Donald W. Hanson, "Is Soviet Strategic Doctrine Superior?" *International Security,* vol. 7, no. 3, Winter 1982-83, pp. 61-83.

6. See Jack L. Snyder, *The Soviet Strategic Culture: Implications for Limited Nuclear Operations,* R-2154-AF (Santa Monica, Calif.: RAND, September

1977); Nathan Leites, *Soviet Style in War* (New York: Crane, Russak, 1982); and Rebecca V. Strode, "Soviet Strategic Style," *Comparative Strategy,* vol. 3, no. 4, 1982, pp. 319-39.

7. The Reagan administration has advertised its strategic thinking as reflecting the idea of "prevailing" as a central concept. However, the administration's force development policy is far closer to the idea of the "countervailing strategy" than it is to the notion of the "prevailing strategy" as defined here. Secretary of Defense Caspar Weinberger explained his view of "prevailing" in a public letter sent to seventy newspaper editors in the United States and abroad. See "U.S. Must Maintain Nuclear Deterrent," *Philadelphia Inquirer,* August 29, 1982; reprinted in *Current News* (USAF), August 29, 1982. p. 1-E.

8. Desmond Ball has argued that the combination of inferior attack-assessment systems and massive co-location of civilian and military assets must render problematical the Soviet ability to distinguish a constrained large counterforce attack from a large countereconomic assault ("U.S. Strategic Forces: How Would They Be Used?" *International Security,* vol. 7, no. 3, Winter 1982-83, pp. 40-41). The claim regarding the technical inferiority of Soviet attack assessment is distinctly challengeable. However, the important point is that it is unlikely that either superpower would be able to appreciate intended restraint in targeting with respect to large weapon laydowns. It is only prudent to assume that attack-assessment capability would be an early victim in a large-scale nuclear war. The utility of an enduring precise attack-assessment capability for the National Command Authorities is a matter of policy consideration that is undecided.

9. Deterrence theorists of the societal-punishment persuasion have long been distressed by the tensions between the demands of strategic flexibility and the requirements of a strategy that should provide only for the punishment of society.

10. Critics of enduring survivability for U.S. strategic assets tend not to challenge directly the argument that endurance is a quality that strengthens deterrence. They do argue that endurance is expensive and can (though logically it should not) endanger the goal of providing sufficient firepower and support needed for the United States to survive the first shock of battle. The critical thesis holds that one might lose a very short war as a result of devoting funds to extensive preparation for a long war.

11. See Desmond Ball, "Soviet ICBM Deployment," *Survival,* vol. 22, no. 4, July-August 1980, pp. 167-70.

12. Ball, "U.S. Strategic Forces," p. 55.

13. Brown, *Department of Defense Annual Report 1981*, p. 65.

14. The most vigorous critique of purported Reagan administration views is Robert Scheer, *With Enough Shovels: Reagan, Bush, and Nuclear War* (New York: Random House, 1982). There has been near constant criticism of the allegedly "first strike" direction claimed to be inherent in the MX Peacekeeper and Trident D-5 programs. For example, see George C. Wilson, "New U.S. Weapons Are Raising Nuclear Fears," *Washington Post,* December 17, 1982. pp. A1, A13.

15. See the testimony of Secretary of Defense Caspar Weinberger in U.S. Senate, Committee on Foreign Relations, *U.S. Strategic Doctrine, Hearing,* 97th Cong. 2d sess., December 14, 1982, p. 12.

16. Scholars of Russian history have noticed that major setbacks for Russian arms triggered political-social upheaval. Imperial Russia appeared not to be resilient in the face of adversity. The examples most often cited are the connections between defeat in the Crimea and the emancipation of the serfs, defeat in the war with Japan and the Revolution of 1905, and of course the consequences of World War I. American analysts have speculated that the Soviet system at home and abroad might not be able to sustain the shock of absence of success—if not of unmistakable defeat. A leading British military analyst has argued, "Soviet military doctrine holds that if war breaks out in Europe, it must be won very quickly by the Soviet Union if it is to be won at all. If the war drags on, there is a high risk that . . . the strain of war will destroy the Soviet Bloc from the inside." C. N. Donnelly, "The Soviet Operational Manoeuvre Group: A New Challenge for NATO," *International Defense Review,* vol. 15, no. 9, 1982, p. 1177. Also highly relevant are Richard Pipes, "Militarism and the Soviet State," *Daedalus,* vol. 109, no. 4, Fall 1980, pp. 1-17; Steve F. Kime, "Warsaw Pact: Juggernaut or Paper Tiger?" *Air Force Magazine,* vol. 65, no. 6, June 1982, pp. 67-69; and Benjamin S. Lambeth, "Uncertainties for the Soviet War Planner," *International Security,* vol. 7, no. 3, Winter 1982-83, pp. 139-66. I do not dismiss the possibility that Soviet anxiety over the domestic consequences of military setbacks may be a factor that strengthens deterrence, but I am concerned lest Western commentators neglect to recognize the quantity and quality of punishment sustained by Russian society in World War I *before* the czarist system fell, and the resilience of the Soviet system in the face of imminent *defeat* in 1941-42. See Norman Stone, *The Eastern Front, 1914-1917* (London: Hodder & Stoughton, 1975); and John Erickson, *The Road to Stalingrad: Stalin's War with Germany,* vol. 1 (London: Weidenfeld & Nicolson, 1975).

17. For a full exposition of this thesis, see Kenneth Hunt, *The Alliance and Europe: Part II: Defending Europe with Fewer Men,* Adelphi Paper no. 98 (London: IISS, Summer 1973). On the feasibility of conventional success for a NATO that had put its house in order, see Thomas A. Callaghan, Jr.,

"Can Europe Be Defended?" *Policy Review*, no. 24, Spring 1983, pp. 75-80; and John J. Mearsheimer, *Conventional Deterrence* (Ithaca, N.Y.: Cornell University Press, 1983).

18. See Colin S. Gray, "National Style in Strategy: The American Example," *International Security*, vol. 6, no. 2, Fall 1981, pp. 21-47.

19. The President's Commission on Strategic Forces (the Scowcroft Commission) offered the judgment that "there can be no doubt that the very scope of the possible tragedy of modern nuclear war, and the increased destruction made possible even by modern non-nuclear technology, *have changed the nature of war itself*" (*Report of the President's Commission on Strategic Forces*, April 1983, p. 2, emphasis added; hereafter cited as Scowcroft Commission Report).

20. One might add dread and fear to the sense of awe. See Michael Mandelbaum, *The Nuclear Revolution: International Politics Before and After Hiroshima* (Cambridge: Cambridge University Press, 1981), chap. 8; and Jonathan Schell, *The Fate of the Earth* (New York: Alfred A. Knopf, 1982). Schell identifies fear as "a moving force behind the establishment of a new system [of political life] by which every decision was made" (p. 222).

21. Great Britain and its army did not develop the kind of military-professional seriousness necessary for defeat of German arms until 1917 and late 1942 in the First and Second World Wars respectively.

22. There is merit in the argument in Michael Howard, "On Fighting a Nuclear War," *International Security*, vol. 5, no. 4, Spring 1981, pp. 3-17.

23. The "probably" is inserted in recognition of the fact that no one can know for certain just what is required for a credible deterrent. The Scowcroft Commission did not doubt the necessity of a war-fighting capability for deterrence. In its report it stated: "The deterrent effect of our strategic forces is not something separate and apart from the ability of those forces to be used against the tools with which the Soviet leaders maintain their power. Deterrence, on the contrary, requires military effectiveness" (Scowcroft Commission Report, p. 7).

24. See Fred Charles Iklé, "Can Nuclear Deterrence Last Out the Century?" *Foreign Affairs*, vol. 51, no. 2, January 1973, pp. 267-85; and Daniel Frei, *Risks of Unintentional Nuclear War* (Totowa, N.J.: Allanheld, Osmun, 1983).

25. Nuclear weapons, regardless of the operational strategies devised for their employment, appear so to have raised the level of damage anticipated in

war that a repetition of "crisis slides" of the 1911-14 or 1938-39 kinds is highly improbable. In the years prior to 1914, general European war was expected, even, to a degree, anticipated with enthusiasm. In 1938-39, war was expected, though without widespread enthusiasm (even in Germany). Particularly useful commentaries on crisis behavior include Phil Williams, *Crisis Management: Confrontation and Diplomacy in the Nuclear Age* (New York: John Wiley & Sons, 1976); Richard Smoke, *War: Controlling Escalation* (Cambridge: Harvard University Press, 1977); Coral Bell, "Crisis Diplomacy," in Laurence Martin, ed., *Strategic Thought in the Nuclear Age* (Baltimore: Johns Hopkins University Press, 1979), pp. 157-85; and Richard Ned Lebow, *Between Peace and War: The Nature of International Crisis* (Baltimore: Johns Hopkins University Press, 1981). For a skeptical view of the crisis-management theories of the nuclear age, see Colin S. Gray, *Strategic Studies and Public Policy: The American Experience* (Lexington, Ky.: University Press of Kentucky, 1982), pp. 107-13.

26. The issues in question pertain largely to what are the proximate as opposed to the underlying causes of war.

27. Weinberger, "U.S. Must Maintain Nuclear Deterrent" (see note 8, above). This judgment was repeatedly publicly many times in 1982 and early 1983; for example: "We believe neither side could win a nuclear war." Caspar W. Weinberger, "The MX Is Needed to Preserve Peace," *USA Today,* May 24, 1983, p. 10.

28. See Edward N. Luttwak, "On the Meaning of Victory," *Washington Quarterly,* vol. 5, no. 4, Autumn 1982, pp. 17-24. As Luttwak illustrates, the idea of "victory" has lost respectability in the United States, in part as a consequence of insistent propaganda by the liberal intelligentsia. It is unfashionable.

29. Bernard Brodie, *War and Politics* (New York: Macmillan Co., 1973), chap. 9.

30. This judgment is confirmed in Bernard Brodie, "The Development of Nuclear Strategy," *International Security,* vol. 2, no. 4, Spring 1978, pp. 65-83. The most carefully researched study of "the professional life and times" of Bernard Brodie is Fred Kaplan, *The Wizards of Armageddon* (New York: Simon & Schuster, 1983). See, also, the arguments in Michael Howard, "The Forgotten Dimensions of Strategy," *Foreign Affairs,* vol. 56, no. 5, Summer 1979, pp. 975-86.

31. See note 23, above.

32. Useful treatments are Stanley Seinkiewicz, "Observations on the Impact of Uncertainty in Strategic Analysis," *World Politics,* vol. 32, no. 1, October 1979, pp. 90-110; P. H. Vigor, "Doubts and Difficulties Confronting a Would-be Soviet Attacker," *RUSI Journal,* vol. 125, no. 2, June 1980, pp. 28-32; and Lambeth, "Uncertainties for the Soviet War Planner" (see note 16, above).

Chapter 2

POLICY GUIDANCE, WAR PLANS, AND STRATEGY

Few if any "war plans" in history have had any authority beyond the first clash of battle. Examples are legion, but 1914 stands as the supreme case when all combatant powers found their war-winning formulas thwarted by disagreeable realities.[1] However, there have been many cases in history where strategic planners have been so fixated on what appeared to be a monumentally stressful immediate undertaking that they neglected to consider ahead of time what they would do if opportunity knocked far more rapidly than expected. The Western Allies, for example, planned meticulously to effect their opposed lodgment on the coast of Normandy in 1944, but they could not bring themselves to plan seriously beyond that. The result was that a plan was missing for rapid exploitation of success out of the beachhead.[2] Similarly, Hitler had been embarrassed in May-June 1940 when his handful of *Panzer* divisions produced a precipitately unfolding campaign success out of what had been intended to be only a victorious battle. Other relevant examples include Egypt in October 1973, Iraq in its ongoing war with Iran, and Great Britain over the Falklands in 1982.[3] In the cases of Egypt and Iraq, military victory was squandered by high commands that had not thought through how tactical success could be transformed into strategic success.

Campaign planning which asserts that everything is important tends to produce a dispersion of scarce effort and a confusion over proximate goals that invite defeat. If everything is of high priority, nothing has the highest priority. German strategy in Russia, and for the defense of *Festung Europa,* suffered from this weakness, as did U.S.-Allied strategy in Europe in 1944-45 and U.S. strategy in South Vietnam.[4]

Clausewitz said of war, "Its grammar, indeed, may be its own, but not its logic."[5] In other words, the dynamics of armed conflict are peculiar to itself, but war can have no meaning in and of itself. By extension, strategic planning must have as its guiding purposes not adverse missile payload drawdowns and the like imposed on an enemy, but political defeats averted and gains recovered. Given that combat is threatened and waged for the ends of policy, whatever those ends may be, it must follow that war plans indicate more or less competently how military power would be applied to achieve the goals set by policy.

War is an uncertain enterprise. Nonetheless, planning for war and the actual conduct of war is an art, or a science, that can be taught. (The major European powers in modern history, copying the Prussian *Kriegsakademie* with greater or lesser success, established general staffs and general staff careers.)[6] While detailed plans for military operations must be short-term and are most relevant when one has the initiative, war planning must be guided by a "theory of victory."[7] A division or corps commander may not be able to see beyond tomorrow's battle, but an army group or theater commander must think in campaign terms, while a government and its most senior military advisers must have as their perspective the conduct of the war as a whole. No government can plan a war in detail beyond the first clash of forces, but it should have a good idea of the strengths and weaknesses of the adversary. Geopolitical factors are not immobile in identity or in meaning over time, but they change slowly.

"Everything in war is very simple, but the simplest thing is difficult. The difficulties accumulate and end by producing a kind of friction that is inconceivable unless one has experienced war."[8] Clausewitz suggests, "Friction is the only concept that more or less corresponds to the factors that distinguish real war from war on paper."[9] Many things that can go wrong will go wrong. Unique though they often appear to a contemporary complainant, many sources of confusion in the real world transcend particular cultures and periods of history.

First, policy guidance almost invariably is ambiguous to a degree. Typically, governments issue instructions to their generals which are phrased so vaguely that while they seem to permit a great deal of local initiative over means they also serve to allow governments, which ultimately are responsible for the conduct of military operations, to assign blame for failure to the military instrument. There are many occasions when distant and inexpert politicians are wise to offer only the most general kind of marching orders to their generals, but there is a tendency to use words to obscure the absence of needed, rigorous thought. For a contemporary example of seemingly sophisticated phrasing covering up a possible vacuum in policy thought, consider the following gem:

Should deterrence fail, our strategy is to *restore peace on favorable terms.* In responding to an enemy attack, we must defeat the attack and achieve our national objectives while limiting—to the extent possible and practicable—the scope of the conflict. We would seek to deny the enemy his political and military goals and to counterattack with sufficient strength to terminate hostilities at the lowest possible level of damage to the United States and its allies. [Emphasis in the original.][10]

What does *"restore peace on favorable terms"* mean? In reality, this is a low-key, circumlocutory euphemism for *win.* The truth is that a U.S. defense community fully occupied with budgetary and/or engineering issues has scant time or talent available for contemplation of the purpose of it all. For the government to have something intelligent to say on the subject of restoring peace on favorable terms, it would have to devote considerable time and energy to the question of war aims. As a very general rule it is accurate to say that the U.S. defense community, official and extraofficial, is preoccupied with establishing the requirements of deterrence, devotes modest attention to the issues pertaining to the actual conduct of war, and has little energy left to address questions of political purpose with regard to the prerequisites for war termination. Should deterrence ever fail, or fail to apply, the meaning of the words "restore peace on favorable terms" would come to be of far more than mere academic interest to U.S. military operators. Those who sign documents that function as policy guidance for the military bureaucracy which compiles war plans should never be permitted to forget that "war plans must be directed at achieving some realistic political and military objectives or they will be empty, senseless, and dangerous."[11]

A second difficulty is that the forces purchased may lack the quantity and quality necessary for the plans drafted to implement the policy guidance to offer prospective military success with confidence. Today, for example, U.S. strategic policy guidance posits requirements that are far removed from contemporary technical feasibility. In the words of the Scowcroft Commission,

> we must be able to put at risk those types of Soviet targets—including hardened ones such as military command bunkers and facilities, missile silos, nuclear weapons and other storage, and the rest—which the Soviet leaders have given every indication by their actions they value most, and which constitute their tools of control and power.[12]

Pending the arrival of the MX intercontinental ballistic missile (ICBM) in the inventory, the target set identified here cannot plausibly be placed at prompt risk.

A third source of confusion is that policy guidance will change in the course of a war, as opportunities beckon or vanish and as the scale of costs is revealed. Knowledge of this fact is less than helpful to officials charged with drafting plans and specifying operational requirements on the basis of today's policy guidance.

Fourth, politicians and soldiers will each be compelled on occasion to transcend their professional expertise and function in the other role. Paradoxically, the newer technologies of global communication promise both to make it much easier for a president to play general (or even sergeant), and to make it likely that many military commanders (including junior officers who have control of small but potent operational forces) will have to make decisions that are as much political as they are military.[13]

A fifth difficulty is that both sides will make mistakes in military preparation and execution. Success will attend the arms of the side that is best able to fight effectively despite its mistakes, that can best work around difficulties, and that can best learn how to change its military behavior while under fire.

Sixth, wars conducted among great powers following long periods of peace (or only minor conflict) that have seen multiple revolutions in weapons technology have always had the character of leaps in the dark. Study and policy action are very different activities. The "leap in the dark" aspects of the French revolutionary wars of the 1790s, World War I, World War II, and (hypothetically) a World War III were and are (with reference to future general conflict) all anticipated with considerable clarity by some students of war. Unfortunately the ultimate goal of strategic inquiry cannot be "truth" but rather effectiveness in strategic practice. In the words of Bernard Brodie:

Strategic thinking, or "theory" if one prefers, is nothing if not pragmatic. Strategy is a "how to do it" study, a guide to accomplishing something and doing it efficiently. As in many other branches of politics, the question that matters in strategy is: "Will the idea work? More important, will it be likely to work under the special circumstances under which it will next be tested?"[14]

Careful study of nuclear conflict, both short and protracted, is of little avail if the government either fails to recognize plausible predictions for what they are or decides that it cannot act on the advice it received. Contemporary U.S. students of the problems of enduring deterrence in a protracted (nuclear) conflict should note the following observation made by a scholar of strategy in World War I:

Even more significant than the prevalence of the short-war view was the fact that those who expected a long war did little to prepare for it. Their inaction was due to the widespread assumption that the state of Europe's political and economic development made such preparations either unnecessary or impossible.[15]

Only one among recent presidents, Jimmy Carter, demonstrated a strong personal interest in the details of nuclear-force readiness and execution. An important reason why war plans and the alert process may (not will or must) fail to serve national policy is that in time of crisis a president and his closest advisers (a group that legally must include the secretary of defense as the designated executive agent of the president in the chain of military command, but that may effectively exclude him if he is not politically close to the president) may discover they did not have options they thought they had.[16] By way of historical analogy, the political leaders of Imperial Germany, Russia, and Great Britain in 1914 did not know what detailed options their general staffs had prepared and, more important, what options had not been prepared.[17] The uniformed war planners do their best to give military expression to the policy guidance they receive, but there is often an absence of genuine dialogue between the professional war planners and those at or close to the highest level of government who inspire, sanction, or simply acquiesce in policy guidance.

It is always possible and to some degree inevitable that mobilization for war renders the outbreak of war more likely.[18] A president who wishes to engage in some fine-tuned political crisis management could discover that the mechanical dynamics of military mobilization, as the forces' alert status was raised, would produce a great deal of "noise" that could obscure his intended signals. A president attempting to conduct an acute political crisis may find that he had far less practicable choice with respect to controlling the details of military preparation and movement than he would like.[19]

With reference to the actual conduct of military operations there is what probably must be characterized as an inherent tension between the demands of military efficiency and the likely political needs of a president or National Command Authority more generally. Although the Joint Strategic Target Planning Staff (JSTPS) is charged with planning to wage nuclear war according to the policy guidance it receives from the secretary of defense and the Joint Chiefs of Staff, its perspective tends properly to be that of the optimally efficient military executive rather than the imaginative strategist. Indeed, one suspects that the function of strategic thought is frequently performed not in any deliberate fashion but by default by military planners. In other words, a concept of prewar, and possibly even of wartime-enduring, deterrence is outlined briefly and more or less ambivalently in a document that the president signs—an NSDM (National Security Decision Memorandum) 242, PD (Presidential Directive) 59, or NSDD (National Security Decision Directive) 13—leaving the strategic function to those who must translate such documents for the guidance of the JSTPS via the nuclear weapons employment policy (NUWEP).[20] The NUWEP process may or may not provide useful insight into what senior policymakers have in mind. Strategy should explain how military power, latent as threat or applied as violence or force, will serve the goals of policy. A uniformed "war planning" staff may fail a president

- if it lacks adequate guidance on what the president intends;
- if the detailed guidance it receives either inadvertently or deliberately subverts the president's intentions;

- if the "grammar" of the war, to which Clausewitz referred, does not permit the character of mobilization and execution process that the president has decided would be politically desirable; or
- if the policy guidance is explicit and faithfully reflective of White House intentions, but the forces (and their logistic and C^3I support architecture) that are maintained and programmed for acquisition are insufficient and/or technically or tactically inappropriate for the specified missions.

Some important ideas relevant to the stability of prewar and intrawar deterrence that have enjoyed high-level official support over the past two decades have been resisted with varying degrees of effectiveness by the uniformed executive agents. Every administration from that of John F. Kennedy to the present day has endorsed the ideas of careful control of and flexibility in the conduct of nuclear operations.[21] Admittedly, the policy story has evolved in important ways, both as the strategic balance has changed and as new weapons have altered what could be done and what needed to be done (e.g., the military target structure has changed massively). Military planners and operators are guardians of the capability to fight their forces efficiently and effectively. Their proximate, even overriding, concerns are to ensure that designated targets will be hit and, to backtrack logically, that the U.S. strategic nuclear forces and their necessary C^3I and logistic support structure actually will be available when called on to fight. The bias of the military planner and prospective responsible operator is toward commitment of forces to battle early and on a large scale. The military planner would like to know that he will have all his country's alert-generated forces available for employment in the first minutes of a war. He does not and cannot know what forces, what logistic support, what C^3I assets will be on-line days or weeks into a war.

To be more specific, uniformed war-planners have been exceedingly nervous about the idea that relatively small nuclear strikes might be executed "up front" as part of a political bargaining process. Such strikes could both damage the ability to execute the single integrated operational plan (SIOP)—some important elements of the forces needed to execute the war plan may have been expended in sub-SIOP skirmishing—and trigger a massive Soviet countermilitary, counter-C^3I, and counterpolitical reply that would have a very disruptive effect on SIOP execution.[22] Also, there tends to be military nervousness over the possibility that the U.S. government might devote scarce resources to protracted conflict capability to the degree that the military's prospects for achieving success in a very short conflict might be compromised. Many uniformed analysts are not opposed to the argument that a protracted conflict capability is desirable for prewar and intrawar deterrence, but they appear to believe that such a capability is a luxury in the context of major shortfalls in U.S. ability to fight well in and survive through a short war.

The desirability of the United States' having strategic forces that can endure in a high-alert or "generated" state for a period of days and weeks—ready to fight on very short notice—has been recognized for many years. The most stressful case of sustained readiness would be a condition of protracted war. All too frequently civilian commentators forget that no matter what U.S. policy guidance and targeting plans prescribe with regard to flexibility, control, and force withholding, armed forces lose efficiency if there is an attempt to hold them in a state of very high readiness. The nuclear-powered ballistic missile submarine (SSBN) force, for example, can be "surged" in time of crisis, but few of the boats will be stocked fully in such an event, and a noticeable fraction of the force will have an endurance that is severely limited by the low level of consumables available on board. The fact that the U.S. SSBN force typically conducts ninety-day patrols and could extend that period for a week or two if need be says little about the prospective operational endurance of SSBNs in time of severe crisis war.

Similarly, there are serious questions about the sustainable readiness of the manned bomber/cruise missile carrier force. If one ignores the most stressful and probably least likely event of a surprise submarine-launched ballistic missile (SLBM) attack on U.S. Strategic Air Command (SAC) airfields (true tactical surprise), one can demonstrate the high probability of SAC flushing aloft an impressively high percentage of its aircraft. But what happens to the scale and character of SAC readiness if the United States either hovers on the brink of war for weeks and months or wages a war confined below the level of a central homeland-to-homeland exchange? In those events men and machines become very tired and, in the latter case, important SAC assets of all kinds would almost certainly be diverted for pressing tactical missions. For example, SAC aerial tankers would be used to support tactical air operations of all kinds, and some bombers would likely be employed in a sea-control role.

The point of this discussion is to illustrate a few of the real-world difficulties that tend to escape careful attention when debate is focused on strategic theory and policy. Forces that would be very efficient in executing major SIOP options within a matter of hours of the beginning of a central nuclear war may suffer quite dramatic drops in efficiency if they must function for days, weeks, or months as an enduring deterrent.

In contrast to the typical soldier's concern that a war be waged efficiently, the politician's concern tends to be that they not be compelled by a politically undisciplined onrush of military events to wage more war than they wish or truly need. Military men tend to assume that if given the choice a politician will opt to fire nuclear weapons "light and late" rather than "heavy and early." In practice, both in actual crises and in the simulated crises of peacetime exercises, there is little evidence suggesting a proclivity on the part of civilians to assume major military risks as the price for retaining diplomatic freedom of action.[23] Nonetheless, the suspicion endures.

From time to time U.S. policymakers and analysts are impressed with the difficulty of providing military plans that would actually be useful in time of need. Somehow the real world rarely seems to cooperate by providing the right enemy in the right place at the right time, so planners seek to provide what flexibility they can, consistent with criteria of military efficiency and with the need to protect the essential integrity of the SIOP as a unified design for force application. Nonetheless, both history and common sense suggest:

The situations in which the use of nuclear weapons is most likely to be initiated are ones which lie *outside* the purview of the contingency plans in the SIOP. Further, the first use of the SIOP forces is likely to follow a period of large-scale military action in which there has already been substantial use of tactical nuclear weapons, significant military and collateral casualties, and some degradation of command and control systems. The dynamics of the escalation process, once set in motion, are likely to foreclose the possibility of employing most of the LNOs [limited nuclear options] and SAOs [selective attack options] in the SIOP.[24]

The logical implication of the condition described here could be a requirement that the U.S. invest in C^3I systems to provide that true "escalation agility" of which Richard Burt has written.[25] However, this may be a case where a plan (within reason) is better than no pre-prepared plan, because it seems to this author unlikely that a U.S. government could or should attempt to design a strategic force application plan in real-time as military and political circumstances evolve. The intellectual trend away from SIOP-like thinking, for fear of the United States' being the captive of an inflexible war plan (after the fashion of Germany in 1914),[26] is healthy and in general sensible, but it founders rapidly on the technical problems of system survivability and the policy problem of the ability of a group of people operating under extreme stress (not to mention imperfect information) to solve tactical and strategic problems on the spur of the moment. Without denying the merit in the ideas of flexibility and "escalation agility," this author is skeptical that a truly major degree of tactical flexibility in potential force application is feasible.

It is ironical that for ten years the U.S. government has been asking for more and more "options" for ever greater flexibility in the potential employment of the strategic forces, at the same time that the prospective operational difficulty of fine-tuning strategic warfare has come to be appreciated very starkly indeed. In the name of flexibility, more and more discriminating strike options have been demanded, even though it is anticipated that the U.S. defense community will never have, pre-planned, exactly the right strategic nuclear option for the unique political circumstance that may arise. Moreover, a strategic doctrine that emphasizes endurance and real-time responsiveness to possibly novel strike demands is a doctrine that stresses prospective technical accomplishment precisely where it is least convincing—with regard to C^3I.

It should not be supposed that the author is strongly pessimistic concerning the feasibility of achieving a tolerable match between policy guidance and military performance. However, the author is concerned lest "policy debate" on strategic nuclear forces subsume consideration and understanding of what is feasible (with reasonably high confidence) in the real world of military organizations, weapons, and their necessary support systems.

Unlike the Soviet Union, with its highly professional general staff system, the United States does not have an authoritative theory of (nuclear) war. The critical question "How do we fight the war if deterrence fails?" occasionally has proved to be a showstopper for government officials and politicians, who would prefer to be able to identify strategic "sufficiency" according to the marginal cost curve that plots destructive energy laid down against industry and population destroyed. But no U.S. government, no matter what its innermost convictions, has persisted in publicly pressing the case for a "war-fighting" doctrine of deterrence.[27] The strategic *and deterrent* logic of the "war-fighting" thesis is inexorable and unequivocal. There must be a balanced strategic posture, embracing complementary offensive and defensive elements, dedicated synergistically to the mission of limiting damage to North America. It is a fact of U.S. (and NATO-European) political culture that professional war-planners, no matter how efficient their offensive-force plans, are planning for wars that the United States could not afford to wage. Any discussion of a potential new offensive-force capability must at least note that it, and its complementary siblings in the triad, are burdened with the need to carry a deterrence story they cannot manage plausibly so long as they are hindered by the absence of U.S. homeland defenses.

It was no accident, as Soviet spokesmen are wont to say, that President Reagan, in his speech on March 23, 1983, in which he offered the first presidential endorsement of the idea of strategic defense to be provided in more than twenty years, did not argue or assert a deterrence or damage-limitation case for such a defense:

I call upon the scientific community who gave us nuclear weapons to turn their great talents to the cause of mankind and world peace: to give us the means of rendering these nuclear weapons [ballistic missiles with nuclear warheads] impotent and obsolete.[28]

Comprehensive endorsement of the idea of damage limitation would have required the president to approve civil defense and air defense as well as ballistic missile defense. President Reagan is reported to have decided that the United States should place less reliance on strategic nuclear *offensive* forces for deterrence in the future; he has directed that studies be undertaken of the implications of ballistic missile defenses for U.S. national security policy.[29] At the present time, however, there is considerable uncertainty over the direction of U.S. strategic nuclear doctrine. Notwithstanding the speech of March 23 and associated decisions and study directives, this author cannot detect as yet a marked transformation in official doctrine in the direction of damage limitation (for deterrence). Not only did the president decide not to set his strategic defense proposal of March 23 in terms of damage limitation, but neither Defense Secretary Weinberger's annual report for fiscal year 1983,[30] nor even the Scowcroft Commission, for all its endorsement of the idea that "deterrence . . . requires military effectiveness,"[31] lends support to the proposition that the U.S. government has recognized the strategic value of the defense of the homeland. This impression is supported strongly by the true test of what is policy and what is not—namely, the test of programs funded.

Strategic argument is not typical of U.S. national security debate. As John Collins has pointed out, the U.S. defense bureaucracy, civilian and uniformed, is not professional in its consideration of the strategic functions of military forces.[32] More often than not, the role of strategic argument in program advocacy is not so much to persuade skeptics as to neutralize the strategic claims of program detractors. Few people are in the habit of thinking strategically about armed forces in general, and fewer still think strategically about nuclear-armed forces.[33] Strategic thinking traditionally has been alien to U.S. policymakers. Until 1943-44, when the preponderance of U.S. material contribution to the Grand Alliance mandated *American* strategic decision-making, the United States appeared to have little need of independent strategic vision. For reasons of geostrategic accident, the United Kingdom in effect performed the strategic determination function for the United States.[34] Moreover, after 1945 it was not really until the 1970s that the United States was acutely in need of genuinely strategic thinking concerning the political value of particular ideas for prospective employment of nuclear weapons.

As was true for the 1950s and 1960s, a preponderance of means rendered *strategic* thought on nuclear threat and use desirable rather than truly essential.

States tend not to develop policy-skill attributes that they do not need. The Eastern Roman, later Byzantine, Empire was supremely strategic in the quality of its statecraft, as (generally though far from invariably) was the Great Britain of the balance-of-power Europe of 1648-1914. Necessity breeds strategic skill. States that are potentially fatally deficient in the physical means of self-defense and exposed geographically to total threats to their security either learn to outthink their stronger adversaries or go under. For most of its history, the United States has had no need to outthink its actual or potential enemies—it could outproduce and if need be outfight them.

Beginning with the great antiballistic missile (ABM) debate of 1969-70, a major public relations dimension has been injected into the strategic-nuclear weapon acquisition process. Unrealistic though it is in historical perspective and with regard to the logic of deterrence, no defense program (be it MX, ABM, or civil defense) can be justified politically on the grounds that it would contribute usefully to the United States' being able to wage, and prevail in, a nuclear war at tolerable cost. The only politically legitimate rationale for nuclear weapon systems is deterrence, but the general public and many opinion leaders of that public seem not to understand that societal punishment is only one approach to deterrence, that it is not synonymous with deterrence.[35]

At least in recent decades, for reasons of political culture and governmental organization the United States has not managed to conduct itself with respect to an arms control process in such a way that the goals of U.S. strategic planners (which may be summarized, at the risk of oversimplification, by the concept of "stable deterrence") would likely be advanced by the negotiating process.[36] This is a sweeping and serious judgment, but it is offered not for the purpose of indictment but rather as an enduring fact of strategic life which U.S. strategic planners and weapon designers ignore at the peril of future acute frustration. Suffice it to say here that U.S. strategic planners, while they should not seek to insist, let alone expect, that strategic arms reduction talks (START) help them achieve the goals they have been directed to pursue, they do have a right, even a duty, to attempt to insist that START should not hurt prospects for the achievement and maintenance of strategic stability. In this damage-limiting perspective, there are plainly certain kinds of agreement that should not be sought. To be specific, so that it not hinder the quest for strategic stability, START should not limit or restrict technologies or tactics that contribute to the survivability of forces.

In 1980, in the immediate wake of the effective political demise of the SALT II treaty in the U.S. Senate, this author seriously entertained the idea that the United States might be willing to turn its back on the arms control practice of the 1970s and reconsider the functions, opportunities, and problems of the multifaceted East-West arms control process. Moreover, and of far more practical significance, this author chose to believe that the American people heard then-candidate Ronald Reagan when he fulminated against the arms control sins of the 1970s.

Today it is evident that arms control considerations enjoy a dominant position over strategic policy scarcely less obviously than was the case through most of the 1970s, at least with respect to declaratory policy and hence with regard to public expectations. There is an important difference between today and the late 1970s in that the relevant policymakers of the Reagan administration know what cannot be achieved through arms control.[37] But policy action, if only in the form of words, is more important than instinct and knowledge. The truth is that the U.S. government has yet to find a way to accommodate the yearning of the American people for "peace" with an effective policy of diplomatic engagement with a Soviet state that views arms control diplomacy as a branch of political warfare. In its report, the Scowcroft Commission achieved a politically effective argument postulating that the quality of forces it proposed would function to provide both effective national security in a unilateral fashion and the critically needed incentives for negotiable arms reduction on a major scale.[38] The near-term political merits of the Scowcroft Commission Report were self-evident; less self-evident were its strategic merits. This subject is addressed in chapter 5.

NOTES

1. A good introduction to this subject is the collection of essays in Paul Kennedy, ed., *The War Plans of the Great Powers, 1880-1914* (London: Allen & Unwin, 1979).

2. This story is well told in Russell F. Weigley, *Eisenhower's Lieutenants: The Campaign of France and Germany, 1944-1945* (Bloomington: Indiana University Press, 1981).

3. See the best of the books available on the Falklands campaign: Max Hastings and Simon Jenkins, *The Battle for the Falklands* (London: Michael Joseph, 1983). The British "war cabinet" was briefed on the plans for the landing at San Carlos on April 27, 1982, but "the only lacuna in the presentation, it was later recalled, was the absence of any discussion of what the land forces would do once ashore" (p. 185). Hastings and Jenkins observe that the San Carlos D-Day plan was "a plan for a landing, not a land campaign" (p. 255).

4. On the Vietnam case, see Harry G. Summers, Jr., *On Strategy: A Critical Analysis of the Vietnam War* (Novato, Calif.: Presidio Press, 1982), chaps. 9, 11.

5. Karl von Clausewitz, *On War*, ed. and trans. Michael Howard and Peter Paret (Princeton: Princeton University Press, 1976), p. 605.

6. See John M. Collins, *U.S. Defense Planning: A Critique* (Boulder, Colo.: Westview Press, 1982), pp. 54-56.

7. I made this argument in detail in "Nuclear Strategy: The Case for a Theory of Victory," *International Security*, vol. 4, no. 1, Summer 1979, pp. 54-87. With respect to Vietnam, Col. Harry Summers wrote, "One thing we did not 'intend to achieve' was victory. . . . Our doctrine specifically excluded it as an aim in war. Testifying before the Senate in 1966, General Maxwell Taylor said that we were not trying to 'defeat' North Vietnam, 'only to cause them to mend their ways' " (*On Strategy*, p. 103).

8. Clausewitz, *On War*, p. 119.

9. Ibid.

10. Caspar W. Weinberger, Secretary of Defense, *Annual Report to the Congress, Fiscal Year 1984*, February 1, 1983, p. 32.

11. Aaron L. Friedberg, "The Evolution of U.S. Strategic 'Doctrine'—1945 to 1981," in Samuel P. Huntington, ed., *The Strategic Imperative: New Policies for American Security* (Cambridge, Mass.: Ballinger Publishing Co. 1982), p. 63.

12. Scowcroft Commission Report, p. 6.

13. In practice, with respect to general war, it is more likely that military men will have to assume the role of politician than that politicians will have to assume the role of general (or sergeant). It must be assumed that prompt and reliable global communications will be unavailable in the event of global conflict. This author is not convinced that *some* communication networks will not survive, but he deems it prudent to be pessimistic. See Desmond Ball, *Can Nuclear War Be Controlled?* Adelphi Paper no. 169 (London: IISS, Autumn 1981); U.S. Congress, Congressional Budget Office, *Strategic Command, Control, and Communications: Alternative Approaches to Modernization* (Washington, October 1981); and William J. Perry, "Technological Prospects," in Barry M. Blechman, ed., *Rethinking the U.S. Strategic Posture* (Cambridge, Mass.: Ballinger Publishing Co., 1982), esp. pp. 145-51; and Paul Bracken, *The Command and Control of Nuclear Forces* (New Haven, Ct.: Yale University Press, 1983).

14. Bernard Brodie, *War and Politics* (New York: Macmillan Co., 1973), p. 452.

15. L. L. Farrar, Jr., *The Short-War Illusion: German Policy, Strategy and Domestic Affairs, August-December 1914* (Santa Barbara, Calif.: ABC-Clio Press, 1973), p. 4.

16. Thomas Powers has written: "The first SIOP was officially approved in December of 1960, but when Nixon came to office, in January of 1969, no President had even been briefed on a SIOP. Indeed the men who wrote the plan, out at Offutt Air Force Base, had to *guess* what sort of nuclear war Presidents wanted to fight, with nothing much solider by way of evidence to go on than official speeches from the President and the Secretary of Defense" (Thomas Powers, "Choosing a Strategy for World War III," *The Atlantic,* November 1982, p. 95). Even though the notion of flexibility is in theory meritorious, no matter how many strategic-force employment options are designed, the list of theoretical options will never match real-world perceived need. The trend of the past decade toward more flexibility in strategic-force applicability reflects contemporary intellectual fashion far more than a sober analysis of likely operational need. An excess of options could help paralyze, rather than liberate, decision.

17. For a classic example of how a nation can secure foreign military obligations without also securing some measure of influence over the plans of allies,

consider the case of Great Britain in 1914. Courtesy of military staff conversations conducted solely at military initiative, the British planned to add a token expeditionary force to the left flank of the French line (though not the far left, because the French, with good reason, did not trust a British army on the coast). But the absence of a formal British commitment and the truly joint military planning to which such a relationship should have led meant that British forces were committed to participate in the execution of the French war plan (Plan XVII), the details of which were unknown to them. See John Terraine, *Douglas Haig: The Educated Soldier* (London: Hutchinson, 1963), chaps. 4-5, esp. p. 62; and S. R. Williamson, *The Politics of Grand Strategy: Britain and France Prepare for War, 1904-14* (Cambridge: Harvard University Press, 1969).

18. Forces in a high-alert status may be likened to a runner on his starting blocks—a false start is always possible when the adrenalin is flowing.

19. Military efficiency and crisis-*management* may have a relation of tension. Acutely aware that mobilization moves might be perceived as steps preparatory to a surprise attack rather than as prudential preparation or as gestures of resolve, a president who wished to cool the military-move fuel to a crisis would have to intervene to prevent the timely and orderly military mobilization process. Surprise-attack issues are well handled in Richard K. Betts, *Surprise Attack: Lessons for Defense Planning* (Washington: Brookings Institution, 1982); and in P. H. Vigor, *Soviet Blitzkrieg Theory* (New York: St. Martin's Press, 1983).

20. On the quality of military planning, and the planning process in general in the United States today, see Collins, *U.S. Defense Planning*. On SIOP development, see Desmond Ball, *Targeting for Strategic Deterrence,* Adelphi Paper no. 185 (London: IISS, August 1981), pp. 106-16, and "U.S. Strategic Forces: How Would They Be Used?" *International Security*, vol. 7, no. 3, Winter 1982-83, pp. 31-60.

21. See Henry S. Rowen, "The Evolution of Strategic Nuclear Doctrine," in Laurence Martin, ed., *Strategic Thought in the Nuclear Age* (Baltimore: Johns Hopkins University Press, 1979), pp. 131-56; and Friedberg, "The Evolution of U.S. Strategic 'Doctrine'—1945 to 1981."

22. The possibility is discussed persuasively in Paul Bracken and Martin Shubik, *Strategic War: What Are the Questions and Who Should Ask Them?* Working Paper no. 50 (New Haven: Yale School of Organization and Management, April 1982). See, also, Bracken, *The Command and Control of Nuclear Forces,* pp. 232-37.

23. See Richard K. Betts, *Soldiers, Statesmen, and Cold War Crises* (Cambridge: Harvard University Press, 1977).

24. Ball, "U.S. Strategic Forces," p. 44.

25. Richard Burt, "Reassessing the Strategic Balance," *International Security*, vol. 5, no. 1, Summer 1980, p. 51.

26. See Gerhard Ritter, *The Schlieffen Plan: Critique of a Myth* (London: Oswald Wolff, 1958); and Barbara W. Tuchman, *The Guns of August—August 1914* (London: Constable, 1962).

27. Secretary of Defense Robert McNamara, for example, in the second half of 1962 supported publicly for only about six months the idea of targeting "withholds" on Soviet cities. However, the SIOP structure, which reflected the "no-cities" targeting idea, introduced as SIOP-63 in July 1962, allegedly was unchanged in important respects for more than ten years. See Ball, "U.S. Strategic Forces," p. 34. McNamara discovered what Defense Secretary Weinberger was to discover in his turn two decades later, that the U.S. government has enormous and possibly insuperable difficulty coping with the public debate that "war-fighting" theories of deterrence tend to trigger and fuel. For obvious reasons of security classification, the evolution of targeting policy is difficult to document. Although this author would register many dissenting opinions with regard to some of the judgments exercised, he recommends the following works as unusually valuable sources of information, roughly in order of historical coverage: David Alan Rosenberg, "The Origins of Overkill: Nuclear Weapons and American Strategy, 1945-1960," *International Security*, vol. 7, no. 4, Spring 1983, pp. 3-71; Desmond Ball, *Politics and Force Levels: The Strategic Missile Program of the Kennedy Administration* (Berkeley: University of California Press, 1980); Desmond Ball, *Déjà Vu: The Return to Counterforce in the Nixon Administration* (Santa Monica, Calif.: California Seminar on Arms Control and Foreign Policy, 1974); Desmond Ball, *Developments in U.S. Strategic Nuclear Policy Under the Carter Administration*, ACIS Working Paper no. 21 (Los Angeles: Center for International and Strategic Affairs, UCLA, February 1980); Ball, *Targeting for Strategic Deterrence;* Powers, "Choosing a Strategy for World War III"; and Fred Kaplan, *The Wizards of Armageddon* (New York: Simon & Schuster, 1983).

28. "President's Speech on Military Spending and a New Defense," quoted in *New York Times,* March 24, 1983, p. 20.

29. These studies, on the technological and policy problems and opportunities in the field of strategic defense, were conducted in the summer and fall of 1983.

30. Weinberger, *Annual Report to the Congress, Fiscal Year 1984;* see pp. 51-58, which are remarkable largely for what they do not say.

31. Scowcroft Commission Report, p. 7.

32. Collins, *U.S. Defense Planning*. By extension, some critics of U.S. defense performance claim that the U.S. armed forces do not prepare adequately for war itself. See Martin Van Creveld, *Fighting Power: German and U.S. Army Performance, 1939-1945* (Westport, Conn.: Greenwood Press, 1982).

33. Many commentators who are not hostile to the very concept of nuclear *strategy* are nonetheless so skeptical of the possibility of actually employing nuclear forces in any controlled, politically purposive manner that their positive contribution to strategic nuclear policy tends to be focused almost exclusively on anxiety manipulation for prewar deterrent effect. Bernard Brodie, Michael Howard, Desmond Ball, and many others say wise things, but they do not say anything very useful to the officials who must provide policy guidance. See the exchange "Correspondence: Perspectives on Fighting Nuclear War" between Colin S. Gray and Michael Howard in *International Security*, vol. 6, no. 1, Summer 1981, pp. 185-87.

34. See Edward N. Luttwak, "On the Meaning of Strategy . . . for the United States in the 1980's," in W. Scott Thompson, ed., *National Security in the 1980's: From Weakness to Strength* (San Francisco: Institute for Contemporary Studies, 1980), esp. pp. 260-63.

35. For a classic example of a misunderstanding of the structure of deterrence theory, see "In Defense of Deterrence" (editorial), *The New Republic*, December 20, 1982, pp. 7-11. What the editors of *The New Republic* were defending was one theory of deterrence—deterrence by the threat to punish Soviet society.

36. See Richard Burt, "The Relevance of Arms Control in the 1980s," *Daedalus*, vol. 110, no. 1, Winter 1981, pp. 159-77. Also of interest are Joseph S. Nye, Jr., "Restarting Arms Control," *Foreign Policy*, no. 47, Summer 1982, pp. 98-113; John Steinbruner, "Fears of War, Programs for Peace," *The Brookings Review*, vol. 1, no. 1, Fall 1982, pp. 6-10; and Colin S. Gray, *Arms Control: Problems*, Information Series no. 132 (Fairfax, Va.: National Institute for Public Policy, January 1983).

37. It is ironical that the policymaking and policy-executing "principals" of the Reagan administration during 1981 and 1982 included the most articulate and cogent of the critics of 1970s-style arms control. Those principals included: Paul Nitze, Eugene Rostow, Edward Rowny, Fred Iklé, Richard Perle, Richard Burt, Richard Pipes, and John Lehman.

38. Scowcroft Commission Report, pp. 22-25.

Chapter 3

NUCLEAR WEAPON POLICY: EVOLUTION AND DEBATE

The nuclear weapon policy of the U.S. government today is mature, robust as far as it goes, bipartisan in its adherents, and has major deficiencies that are not particularly controversial at the level of simple itemization. To be specific, the United States does not have in its nuclear arsenal anything close to the optimal mix of weapons of different kinds, or of C^3I and logistic support assets, for the efficient execution of the chosen strategy design; and as observed in chapter 2, the defensive side (both active and passive) of the strategic posture continues to languish. However, over the past few years a great deal has been accomplished intellectually concerning what might be termed the conceptual basis for strategic policy architecture. Basic questions have been posed, and persuasive-looking answers have been approved by two administrations of different political parties.

It is a commonplace to observe that, contrary to textbook teaching, weapons drive strategy rather than vice versa, it is alleged that war planners plan to do what weapon developers permit them to do. Although one can do only what weapons permit, one might also attempt to implement a strategic design with weapons that are not well suited for the purpose. That is the situation in the United States today. Contemporary U.S. strategic nuclear forces are composed of an inventory of weapons by and large designed in the 1960s or earlier to suit a targeting philosophy substantially different from that which has evolved over the past decade. Similarly, the C^3I support today for U.S. SIOP and post-SIOP forces was "designed" or more accurately grew in an era where far less rigorous demands had to be met.[1] There can be no doubt that in the early 1980s strategy is in the driver's seat. This is as it should be, but it does serve to expose the government of the day to the charge that it is talking about and planning for military operations in the absence of an appropriate mix of forces. In addition, if strategic vision is critically dependent on technological prowess yet to be demonstrated, it invites the charge that the visionary does not know whether his vision is technically feasible.[2] Given that strategic weapons typically consume ten years or more from initial blueprints to substantial operational capability, and that strategic ideas for the revision of policy guidance to targeteers can be altered in a matter of weeks or months, the problems of mismatch among declaratory policy, policy guidance, war plans, and forces are perennial.[3]

The idea that politicians and officials are little more than puppets in the hands of the developers and prospective users of military hardware is one of the plausible fallacies of history.[4] The deficiencies in U.S. national security policy today are not so much in the realm of intellectual guidance as they are at the sharp end of available forces and C^3I with suitable technical characteristics. While it is true that one can devise strategies only for employment of weapons that either exist or are expected with high confidence to exist in the near future, it is not right to assert, as Thomas Powers does, that "shifting to 'realistic' planning for limited war, so alarming to the general public, was not anything Carter or his advisers *chose* to do. They were pushed every step of the way by the weapons themselves." (Emphasis in original.)[5]

Present U.S. nuclear strategy—at least at the high level of policy guidance (as opposed to the operational level of implementation of that guidance in practice by the Joint Strategic Target Planning Staff—is about as far removed from the implications of Powers's mistaken judgment as possible. It is no exaggeration to say that the United States has been compelled in recent years by the adverse trend in the multilevel East-West military balance to think strategically about nuclear weapons virtually for the first time in a generation. Given the societal and political constraints in a democracy which tend to inhibit realistic thought in peacetime on the subject of war—albeit for the proximate purpose of war prevention or deterrence—it is surprising that U.S. nuclear strategy is as oriented toward military effectiveness as it is.

Thomas Powers and others who discern technological determinism err most seriously in their confusion of the possible with the necessary. Flexibility in nuclear targeting and enduring political control of nuclear forces will be possible in the years ahead,[6] but they will be features of U.S. policy not so much because they are possible but because they are judged to be strategically necessary.

Nuclear strategy and its force-postural expression must be relevant to the support of vital U.S. national interest in the following phases of international political life:

- Peacetime competition
- Crisis time
- Wartime
- War termination and postwar competition

Both critics and supporters of Presidential Directive 59 (PD-59, on nuclear weapons employment policy, signed by President Carter on July 25, 1980) have claimed incorrectly that that document signaled a major break in U.S. nuclear strategy. Former Defense Secretary Harold Brown,[7] and later Caspar Weinberger, more accurately stressed the essential continuity in official strategic thinking.[8] It is often claimed that since summer 1980 the United States has shifted from a deterrent approach to nuclear conflict to a war-fighting approach.[9] More accurately, though still erroneously, it is alternatively claimed that the United States has moved its policy from a punitive view of the requirements of deterrence to a war-fighting, or "classical strategy," view of deterrence. In the words of one investigative journalist,

The principal purpose of PD-59, and of the continuing refinements of war-fighting strategies that have followed, is still deterrence, but it is deterrence of the sort that prevailed before 1914.[10]

And,

PD-59, and the new SIOP it led to, put the "war" back in strategic war. It was based on a theory of what mattered most to the Soviet government—military forces, war-related industry, continued control of the satellite countries in Eastern Europe, security from China in the East, and domination of the U.S.S.R. by the Communist Party. In the event that deterrence failed and hostilities broke out, PD-59 proposed a military campaign in the traditional sense, targeted on the immediate sources of an enemy's ability to fight.[11]

This characterization is true as far as it goes, and it is certainly true with respect to the intentions of some of the people most intimately involved in the production and implementation of PD-59. However, it seriously overstates the case. PD-59, and subsequent refinements to nuclear weapons employment policy, does not and cannot afford totally to abandon a punitive approach to deterrence in favor of a "wage and win the war" approach. While PD-59 may have marked a new plateau, if not a "sea change in official attitudes toward nuclear war,"[12] that new plateau had important precursors and was not without significant ambiguities. For example, the U.S. government has not abandoned the idea that the threat of unacceptable punishment may coerce the Soviet Union into applying militarily unwelcome discipline on its strategic forces; and the U.S. government is far from committed to the proposition that in extreme circumstances the United States must have the capability to win a short or protracted (nuclear) war. PD-59 did deemphasize general industrial-economic targeting (believed to be most relevant to postwar recovery) in favor of targeting war-supporting economic activity,[13] but it also is alleged to have made explicit provision for the need to hold general industrial (for which read also civilian population, given the colocation of the two) assets at residual risk. As noted above, Harold Brown balanced his presentation of the countermilitary, or "warfighting," theme in the countervailing strategy, with a strong restatement of the traditional assured destruction requirement vis-à-vis Soviet economic endeavor in general.[14]

Several alternative descriptions of current U.S. nuclear strategy are substantially correct. There is some genuine ambivalence. Moreover, there are both praiseworthy and not-so-praiseworthy reasons for the mix of ideas that comprise that strategy. The paragraphs that follow provide a critical summary of the key ideas in contemporary strategic policy.

First, U.S. nuclear strategy today is founded on the belief that as *the* fundamental requirement the Soviet Union should anticipate defeat in the event of war; this is a "denial of victory" approach.[15] Second, it is believed that victory, in Soviet terms (which are the ones that matter if the U.S. deterrent is to be a deterrent), is impossible if the essential military, paramilitary, and police assets of the state are damaged to the point where they cannot perform their functions and/or if the political control functions of the Soviet state cannot operate at a minimum essential level.[16]

Third, notwithstanding some policy declarations and guidance that have stressed the desirability of the United States' downgrading the role of offensive forces in nuclear strategy, and a fairly clear trend toward greater military effectiveness,[17] U.S. strategic policy still, at root, relies on a punishment strategy. Indeed, U.S. nuclear strategy is a sophisticated variant of the "competition in risk-taking" described so eloquently by Thomas Schelling more than twenty years ago.[18] Today, preeminently, the United States plans to punish the Soviet state and its executive instruments rather than its captive, or acquiescent, society. The official U.S. concept of deterrence depends on the idea that a Soviet leadership, anticipating severe damage to its most cherished values, will be deterred from choosing to fight, or would be deterred from pressing on with a conflict that already had begun.[19]

Fourth, the trend in official thinking today is to focus almost exclusively on Soviet targets of direct or at least very plausible relevance to the waging of the war. To that extent the United States has a "war-fighting" strategy (for deterrence), provided one notes the continuing ambiguity over the downstream threat to more general Soviet economic activity.

Fifth, nuclear strategy today fails the central test for being a true war-fighting, or "classical strategy," approach to conflict in that it offers no physical shield for the United States. In referring to PD-59 as intended to provide "the sort of deterrence that prevailed before 1914," Powers failed to notice this fact. The French army stood between the German army and Paris. The most important strategic difference between the nuclear era and all previous eras in weaponry is that in the nuclear era a state (or society) can be defeated even if its armed forces are not. There is a very imperfect parallel in the campaigns waged by German submarines in World Wars I and II and in the maritime surface blockade by Great Britain in World War I. However, there is no true historical analogue to the principal function of the most potent of U.S. weapons today, which is to visit devastating retaliation on a nation that already may have written *finis* to the American experience.

Sixth, U.S. strategy today posits the requirement that the

United States should be able to prevail in war, whether the war be short or long. The U.S. government has not chosen a protracted war strategy; it has simply observed that the quality of crisis, and intrawar, deterrence should be enhanced usefully if the Soviet leadership is convinced the United States has the means to protract a conflict. The Soviet Union knows that the strain of protracted conflict can be the midwife of revolution. Soviet anxieties over their vulnerability in the event of a lengthy campaign appear to be reflected in their choice of the "operational maneuver group" strategy for war in Europe.[20]

Seventh, official U.S. thinking on nuclear strategy, in its apparent endorsement of the idea of protracted conflict as integral to the architecture of deterrence, requires a theory—and, better still, capabilities—that would enable it to apply force over a period of weeks and perhaps even months. Clearly, protracted conflict, or "enduring deterrence" (for the official euphemism), is a subject that could benefit from a great deal of careful and scholarly attention. Soviet thinking on this subject may have been misunderstood. The Soviet state has been engaged in a protracted conflict with the antagonistic social systems of the West ever since 1917. Although Western politicians, officials, commentators, and scholars have discerned eras of cold war and eras of détente, it is obvious that the Soviet state has been conducting political warfare (or "war in peace") against us in one way or another through all those eras.

Both Soviet history and strategic common sense illustrate that no nation chooses to wage a protracted conflict. Everybody prefers to win quickly. U.S. defense analysts, official and otherwise, noticed in the late 1970s that the Soviet Union seemed not to be preparing only for "Armageddon in an afternoon." The Soviet Union has provided both physical and written evidence of anticipating that a general war might last a matter of weeks or even months, but that evidence seems to reflect Soviet anxieties rather than Soviet wishes or expectations. The U.S. defense community has reasoned that the Soviet political leadership finds the concept of protracted war to be especially deterring. As supremely conservative political realists, Soviet political leaders understand just how tenuous the authority of the Soviet empire (both territorial and hegemonic) might be in a context of real military stress.[21]

The trend in U.S. nuclear strategy toward an as-yet-unconsummated marriage of planned offensive and defensive actions for the ends of deterring undesired events and, if need be, waging a war to a successful conclusion probably should be judged inexorable. The reassertion of the ideas of classical strategy in the Kennedy administration, when the secretary of defense offered the opinion that "nuclear war should be approached in much the same way that more conventional military operations have been approached in the past,"[22] was resisted strongly, but those ideas have always returned with a regularity that reflects neither the persistence of atavistic attitudes nor the baneful influence of sinister would-be nuclear warriors.[23]

The "classical strategy" approach to nuclear weapons, which insists that planners must treat nuclear weapons as weapons and not as instruments of the wrath of God,[24] has endured and grown greatly in strength because it is logically inescapable. The official nuclear-policymaking community cannot purposively plan the Armageddon.[25] Scholars and journalists enjoy the luxury of being able to reject the idea of limited nuclear war on the basis of such elevated reasoning as the following:

Until such time as a persuasive account is offered of the relationship between war's political ends and its nuclear means, the conclusion that must inescapably be drawn is that, as a matter of political theory, the macro-limitations inherent in war itself must serve as a prohibition upon resort to this particular means.[26]

The safest course is for the scholar simply to restate the problem and avoid the vulnerability that comes from advancing a theory of nuclear campaigning of his own. U.S. officials do not need to be reminded that nuclear weapons constitute a two-edged sword. A well-known scholar has offered the following unhelpful pronouncement:

The question of what happens if deterrence fails is vital for the intellectual cohesion and credibility of nuclear strategy. A proper answer requires more than the design of means to wage nuclear war in a wide variety of ways, but something sufficiently plausible to appear as a tolerably rational course of action which has a realistic chance of leading to a satisfactory outcome. It now seems unlikely that such an answer can be found. No operational nuclear strategy has yet to be devised that does not carry an enormous risk of degenerating into a bloody contest of resolve or a furious exchange of devastating and crippling blows against the political and economic centers of the industrialized world.[27]

The author of the above had previously made the strange observation that "if weapons had to be designed for operational use then some sort of guidance was necessary, which required stating a preference for one form of nuclear employment against another."[28] The observation is strange because *all* weapons are "designed for operational use"; they may be ill-designed, or the political purposes may be ill-chosen, but those are other matters.

Lawrence Freedman, author of the above statements, and others who are wont to comment unsympathetically on the fruits of the endeavor of nuclear policymakers to design strategies that might have some operational merit,[29] require that the U.S. government provide "something sufficiently plausible to appear as a tolerably rational course of action which has a realistic chance of leading to a satisfactory outcome." But what would constitute a satisfactory outcome, given the technological facts of the nuclear age? U.S. officials and their advisers have toiled with imagination to find answers to the question "What should we do if deterrence fails?" that have *strategic* merit. Answers have been found, but they entail waging campaigns that must produce millions of U.S. casualties.

Strategic thinkers, be they armchair commentators or responsible officials, should never forget that wars between great powers have always carried "an enormous risk of degenerating into a bloody contest of resolve." Officials should, and do, work to reduce that risk, but they cannot eliminate it. That being the case, what constructive advice can be offered? Lawrence Freedman, Michael Howard, and a host of others tend not to have advice to offer. In a manner akin to that of the "ground zero" movement of spring 1982,[30] they warn us that nuclear war would be terrible[31] and that it is difficult if not impossible to make operational strategic use of nuclear weapons (i.e., actually to employ nuclear force for political purposes). For the better part of two decades the U.S. government has believed that discretion is the better part of valor with regard to public explanation of its nuclear strategy. As a consequence the general public has the impression that strategic stability relates to the certain ability of each superpower to destroy the major cities of the other, that above and beyond this basic capability one is in the region of nuclear overkill. This level of understanding of official U.S. deterrence reasoning translates into major political problems for such weapon systems as the MX (however based) and the Trident D-5, both of which plainly are overdesigned for a city-threatening role.

The evolution of U.S. nuclear strategy since the late 1940s has shown a fairly orderly response to major influencing factors. The strategy story has reflected the impact of the available size and quality of the nuclear arsenal and its delivery vehicles; intelligence concerning and growth in the size of candidate target sets; the growth in the quantity and quality of Soviet strategic forces; and assessment of probable Soviet operational style in war-waging—as a part of U.S. understanding of Soviet political and strategic culture. Additional important factors include the personal impact of key individuals and the competitive play of intragovernmental bureaucratic interests.[32]

Beyond the ranks of defense professionals of both parties, current nuclear policy debate has surfaced what may be termed a radical critique not only of the nuclear strategy of the Reagan administration but also of the nuclear strategy of every administration that has had to have a nuclear strategy. This radical critique is important because it has posed some legitimate and significant questions, because it is having a growing influence on the politics of defense and hence on the ability of the government to carry through its strategic planning proposals, and because it illustrates the extent of the public education task faced by the government. The paragraphs that follow summarize the principal elements of the radical critique.

First, nuclear strategy can have policy utility only in nonuse. It scarcely matters what U.S. operational strategy is, because people are deterred by the prospect of nuclear war per se, not by the anticipation of suffering particular kinds of damage.

Second, notwithstanding the point just made, it is prudent for the United States to develop and sustain an "inoffensive deterrent" posture, one that does not threaten major elements of the Soviet strike-back capability. History shows that folly in high places is always possible.[33] Counterforce capability may appear to promise a sufficient fraction of a "theory of victory" that decisionmakers who should be deterred from very rash actions might not be deterred.

Third, the balance of terror is massively indelicate,[34] in part because deterrent effect flows at least as much from perceptions of "national will and resolve" as from mathematical calculations of net strategic advantage in strategic assets.[35] Many assessments of the stability of deterrence err because they do not take account of the defender's inherent advantage in presumed political will.[36] This proposition posits, for example, that NATO Europe is far more interested in defending itself than the Soviet Union is in adding it to its hegemonic empire. This proposition is a logical truth of some merit, but it lends itself to abuse by an unscrupulous debater. History shows that "defenders" have been intimidated by aggressors regardless of the value imbalance of the conflict in favor of the defender, and this author is not prepared to judge that the Soviet Union will always be less interested in controlling Western Europe, Middle Eastern oil, or whatever than the defending parties will be in thwarting them. If the strategic balance truly is not delicate, it should be difficult for radical critics of defense policy to argue that the "rogue weapon of the moment" is really much of a danger to international peace and security.[37] Radical critics of mainstream thinking on strategic weapon policy tend to oppose every proposed innovation as if each was the potential error that would create a fatal instability in the East-West deterrence system.[38]

The fourth principal element of the radical critique is that there is no credibility-of-deterrence problem that should drive the United States to more and more differentiated nuclear-war plans and to more and more counterforce-capable weapons and C^3I support systems. Politicians dread nuclear war per se, and the possibility of nuclear war lurks in all East-West crises. The sometimes elegant architecture of U.S. deterrence theory has no benign relevance to international peace and stability. Intellectually, though not psychologically or emotionally, rigorous thought about nuclear strategy is as unnecessary for a sufficient deterrence policy as it may be dangerous in its potential to breed contempt as a result of overfamiliarity.[39] An analyst who spends his most productive years refining nuclear arsenal exchange models of great mathematical distinction may just fool himself into believing that he knows something important about nuclear war. More to the point, he may fool some senior officials into believing that he knows something important about nuclear war.[40]

Fifth, the "nominalist fallacy" pervades nuclear policy and escapes recognition for what it is. Specifically, people confuse ideas with reality. To conceive of limited nuclear war, for example, and to be able to describe its necessary prerequisites, desirable rules of engagement, and the like is merely an intellectual pastime. A gifted theoretician explaining the logical architecture of a theory of limited nuclear war is displaying his virtuosity, not revealing any necessary truth about the nature of nuclear combat.[41]

The sixth and final principal element of the radical critique is that official and extraofficial nuclear policy theorists of a NUT (nuclear utilization theory)[42] persuasion allegedly are virtually silent on the central issue of war termination. Official nuclear planning and the public nuclear-policy literature seem to draw a discreet veil over the nuclear-war endgame. If strategy is the bridge between policy and military power, how does the United States conclude nuclear hostilities successfully?[43] Conservative and liberal critics of nuclear policy tend to agree that the absence of a plausible "theory of victory" (or theory of war termination that accommodates U.S. achievement of at least its minimal war aims) would be a fatal weakness in official policy.

There are so many critics of U.S. nuclear policy, with so many different preferences and levels of political, technical, and doctrinal understanding, that truly fair characterization of the contemporary debate is probably not possible. Most of the more radical critiques of extant U.S. nuclear strategy, on the basis of subjective judgment by this particular author, do identify some genuine problems. Regardless of one's doctrinal affiliation, one must recognize prudently that (1) there is massive uncertainty over "what deters" (who? on what issue? when?); (2) irresponsible adventurers very occasionally do secure the commanding heights of authority in great powers; (3) deterrent effect is enhanced both by what we think we know about nuclear war and by fear of the unknown; (4) Clausewitz's idea of "the culminating point of victory" may apply to nuclear policy[44] (in short, it is possible that the stability of mutual deterrence may be endangered if one proceeds too far into the "rational employment" realm); (5) limited nuclear war is as persuasive in theory as it may be implausible in the context of real-world constraining factors (Clausewitz's "friction");[45] and (6) it is as easy to understand why a nuclear war should be terminated promptly for political-strategic reasons as it is difficult to see how it could be so terminated given the damage that the C^3 systems of both sides certainly would have suffered.

The weaknesses apparent today in U.S. nuclear strategy and in the strategic nuclear force posture are fairly obvious and are relatively undisputed. Unfortunately, much of the more fundamental criticism of U.S. nuclear strategy neglects to recognize the existence of a realm of policy necessity. Officials responsible for research and development of a particular controversial weapon option may and should draw some comfort from the fact that although critics of nuclear strategy per se, or of the nuclear strategy that currently is authoritative, do raise many important and legitimate questions, those critics tend to be more persuasive in the realm of destructive questioning than they are when they venture into the field of constructive proposals.

The radical critique of U.S. nuclear strategy neglects the key fact that the strategy has been driven preeminently by the need to support U.S. foreign policy interests.[46] The United States has purchased countermilitary strategic capabilities over the past thirty years because, in the apposite words of Edward N. Luttwak, "superpowers, like other institutions known to us, are in the protection business. When they cannot protect clients, they lose influence, not just locally but worldwide."[47]

For the mental health of American policymakers—not to mention deterrent credibility or ethical considerations of an Augustinian "just war" kind—military means should bear some relationship to military ends. The logical implication of much of the radical critique of extant U.S. strategic-nuclear planning is that the United States should "plan" only to effect a general holocaust. It so happens that human beings do draw an important distinction between what is certain and what is probable. In the context of this discussion, one may agree that it is improbable that strategic-nuclear use could be controlled closely and in a purposive manner by central political authority, but that nonetheless it is essential that one should attempt the strategic application of nuclear force.

NOTES

1. See U.S. Congress, Congressional Budget Office, *Strategic Command, Control, and Communications: Alternative Approaches to Modernization* (Washington, October 1981); Jonathan B. Tucker, "Strategic Command-and-Control Vulnerabilities: Dangers and Remedies," *Orbis,* vol. 26, no. 4, Winter 1983, pp. 941-63; and Paul Bracken, *The Command and Control of Nuclear Forces* (New Haven, Conn.: Yale University Press, 1983).

2. President Reagan attracted this charge with his March 23, 1983, remarks in favor of ballistic missile defense of the U.S. homeland. "President's Speech on Military Spending and a New Defense," *New York Times,* March 24, 1983, p. 20.

3. See Colin S. Gray, *Strategic Studies and Public Policy: The American Experience* (Lexington: University Press of Kentucky, 1982), chap. 2.

4. See Trevor N. Dupuy, *The Evolution of Weapons and Warfare* (New York: Bobbs-Merrill Co., 1980).

5. Thomas Powers, "Choosing a Strategy for World War III," *The Atlantic,* November 1982, p. 84.

6. Which is not to deny that the possible may not be probable. Strong arguments that stress the problems are Desmond Ball, *Can Nuclear War Be Controlled?* Adelphi Paper no. 169 (London: IISS, Autumn 1981); and John D. Steinbruner, "Nuclear Decapitation," *Foreign Policy,* no. 45, Winter 1981-82, pp. 16-28.

7. Harold Brown, Remarks Prepared for Delivery at the Naval War College, Newport, Rhode Island, August 20, 1980.

8. Caspar Weinberger, "U.S. Must Maintain Nuclear Deterrent," *The Philadelphia Inquirer,* August 29, 1982, reprinted in *Current News* (USAF), August 29, 1982, p. 1-E.

9. Secretary of Defense Weinberger has summarized the differences between late-Carter-era and Reagan-era strategic policy in U.S. Congress, Senate, Committee on Foreign Relations, U.S. Strategic Doctrine, Hearing, 97th Cong., 2d sess., December 14, 1982, p. 100. For a useful critical analysis, see Jeffrey Richelson, "PD-59, NSDD-13 and the Reagan Strategic Modernization Program," *The Journal of Strategic Studies,* vol. 6, no. 2, June 1983, pp. 125-46.

10. Powers, "Choosing a Strategy for World War III," p. 109.

11. Ibid., p. 108. Also see Thomas Powers, *Thinking About the Next War* (New York: Alfred A. Knopf, 1982), chap. 11.

12. Powers, "Choosing a Strategy for World War III," p. 103.

13. See Desmond Ball, *Targeting for Strategic Deterrence,* Adelphi Paper no. 185 (London: IISS, Summer 1983), pp. 23, 30-31.

14. See Harold Brown, *Department of Defense Annual Report, Fiscal Year 1981,* January 29, 1980, p. 65.

15. This was an innovation in U.S. policy thinking introduced by the Carter administration. Sensitivity to probable *Soviet* reasoning on *their* incentives and disincentives to fight was not a prominent characteristic of U.S. government thinking on the requirements of deterrence for the first thirty-plus years of the nuclear age. An excellent discussion is Keith B. Payne, *Nuclear Deterrence in U.S.-Soviet Relations* (Boulder, Colo.: Westview Press, 1982).

16. Countercontrol targeting has been heavily criticized for its likely negative impact on Soviet ability to negotiate war termination, even simply to control their forces, and on their motivation to exercise targeting restraint. To date no one has criticized the idea on the grounds that it misses the mark in likely deterrent effect. Countercontrol targeting is not synonymous with ethnic targeting, though the two may be related. I am not sympathetic to ethnic targeting. See Gary L. Guertner, "Strategic Vulnerability of a Multinational State: Deterring the Soviet Union," *Political Science Quarterly,* vol. 96, no. 2, Summer 1981, pp. 209-23; and George H. Quester, "Ethnic Targeting: A Bad Idea Whose Time Has Come," *Journal of Strategic Studies,* September/October 1982, pp. 228-35. Quester is simply wrong. Ethnic targeting does not figure in U.S. nuclear policy today. An interesting critique of U.S. targeting ideas of recent years is Jeffrey T. Richelson, "The Dilemmas of Counterpower Targeting," *Comparative Strategy,* vol. 2, no. 3, 1980, pp. 223-37. Countercontrol targeting is discussed also in Colin S. Gray, "Targeting Problems for Central War," *Naval War College Review,* vol. 33, no. 1, January-February 1980, esp. pp. 12-15; and Ball, *Targeting for Strategic Deterrence,* pp. 31-32.

17. The Scowcroft Commission (in its *Report of the President's Commission on Strategic Forces,* April 1983) makes it plain that in its view there is an over-

riding "firepower," or counterforce, case for deployment of the MX ICBM (pp. 16-18). The offensive "firepower" argument also is deployed as a major rationale for a highly accurate small single-warhead ICBM (p. 15).

18. See Thomas C. Schelling, *The Strategy of Conflict* (Cambridge: Harvard University Press, 1960); and idem, *Arms and Influence* (New Haven: Yale University Press, 1966).

19. Innovative thinking on deterrence is rare, which should be expected of a subject that has accumulated a vast literature over the past thirty years. For refreshing challenges to orthodox thinking, see Samuel P. Huntington, "The Renewal of Strategy," in Huntington, ed., *The Strategic Imperative: New Policies for American Security* (Cambridge, Mass.: Ballinger Publishing Co., 1982), pp. 1-52.

20. See P. H. Vigor, *Soviet Blitzkrieg Theory* (New York: St. Martin's Press, 1983).

21. Indirect evidence of Soviet anxiety over the political resilience of their polity in war is provided by the brutal action taken against Hungary and Czechoslovakia in 1956 and 1968 and much more recently by the indirect action of the Polish armed forces. Also, Soviet official intolerance of individual organized dissidence provides some basis for gauging Soviet official sensitivities. See George W. Simmonds, ed., *Nationalism in the U.S.S.R. and Eastern Europe in the Era of Brezhnev and Kosygin* (Detroit: University of Detroit Press, 1977); Helene Carrere d'Encausse, *Decline of an Empire: The Soviet Socialist Republics in Revolt* (New York: Newsweek, 1979); and Seweryn Bialer, *Stalin's Successors: Leadership, Stability, and Change in the Soviet Union* (Cambridge: Cambridge University Press, 1980).

22. Robert McNamara, quoted in William W. Kaufmann, *The McNamara Strategy* (New York: Harper & Row, 1964), p. 116.

23. The recurring fashionableness of some aspects of classical strategic reasoning has been a feature in *declaratory policy*. Operational strategy for force employment has never veered from a strong commitment to counterforce targeting. Desmond Ball risked misleading his readers when he entitled an important study *Déjà Vu: The Return to Counterforce in the Nixon Administration* (Santa Monica, Calif.: California Seminar on Arms Control and Foreign Policy, December 1974). The argument for the persistence of baneful influence pervades Fred Kaplan, *The Wizards of Armageddon* (New York: Simon & Schuster, 1983). The title tells all.

24. See Payne, *Nuclear Deterrence in U.S.-Soviet Relations*, chap. 8.

25. Even Jonathan Schell does not accuse officials of such a high crime (*The Fate of the Earth* [New York: Alfred A. Knopf, 1982]). Typically the Reagan

administration is criticized from the perspective that "we see that our government has too little fear [of nuclear war]." Letter to the editor by David P. Barash, *Bulletin of the Atomic Scientists,* vol. 39, no. 1, January 1983, p. 52.

26. Ian Clark, *Limited Nuclear War: Political Theory and War Conventions* (Princeton: Princeton University Press, 1982), p. 240.

27. Lawrence Freedman, *The Evolution of Nuclear Strategy* (London: Macmillan & Co., 1982), p. 395.

28. Ibid., p. 393.

29. E.g., Michael Howard, "On Fighting a Nuclear War," *International Security,* vol. 5, no. 4, Spring 1981, pp. 3-17.

30. For the standard text, see Ground Zero, *Nuclear War: What's in It for You?* (New York: Pocket Books, 1982). The "follow-up" work is Ground Zero, *What About the Russians—and Nuclear War?* (New York: Pocket Books, 1983).

31. Arthur M. Katz, *Life After Nuclear War: The Economic and Social Impacts of Nuclear Attacks on the United States* (Cambridge, Mass.: Ballinger Publishing Co., 1982).

32. The best introduction to the targeting policy aspects of nuclear strategy is Ball, *Targeting for Strategic Deterrence.* Also useful are Henry Rowen, "The Evolution of Strategic Nuclear Doctrine," in Laurence Martin, ed., *Strategic Thought in the Nuclear Age* (Baltimore: Johns Hopkins University Press, 1979), pp. 131-56; and Bracken, *The Command and Control of Nuclear Forces,* chap. 3.

33. An ICBM force based deep underground, for example, would be an inoffensive deterrent in that it would have only second (and beyond) strike utility. The posture of NATO's ground forces in Europe constitutes an "inoffensive deterrent" in that the ground forces lack a convincing ground-seizing capability. The Group of Soviet Forces in Germany, whatever the true character of Soviet political intentions, is an excellent example of an offensive deterrent capability. See C. N. Donnelly, "The Soviet Operational Manoeuvre Group: a new challenge for NATO," *International Defense Review,* vol. 15, no. 9, 1982, pp. 1177-86.

34. Contrary to the prudential reasoning in the classic article by Albert Wohlstetter, "The Delicate Balance of Terror," *Foreign Affairs,* vol. 37, no. 2, January 1959, pp. 211-34.

35. The most forthright presentation of this argument is Richard Ned Lebow, "Misconceptions in American Strategic Assessment," *Political Science Quarterly*, vol. 97, no. 2, Summer 1982, pp. 187-206.

36. See David Dessler, " 'Just in case'—the danger of flexible response," *Bulletin of the Atomic Scientists*, vol. 38, no. 9, November 1982, pp. 56-57.

37. Laurence Martin has argued that the arms control process encourages both Western analysts and Soviet diplomats, in the political context in the West of a "prejudice against technological innovation," to engage in "a series of running fights over whatever is designated 'rogue weapon' of the moment by the disarmament lobby or by Soviet diplomacy." *The Two Edged Sword: Armed Force in the Modern World* (London: Weidenfeld & Nicolson, 1982), p. 72.

38. For an empirical check on the argument in the text, skeptical readers can review the voluminous public record of opinions on individual weapon systems registered over the past decade by such people as Herbert Scoville, Jr., Paul C. Warnke, and Jeremy Stone.

39. See Fred Iklé, "Can Nuclear Deterrence Last Out the Century?" *Foreign Affairs*, vol. 51, no. 2, January 1973, esp. pp. 278-82.

40. This thought pervades Kaplan, *The Wizards of Armageddon*.

41. The classic example of this phenomenon is Kahn's *hypothetical* escalation ladder. See Herman Kahn, *On Escalation: Metaphors and Scenarios* (New York: Praeger Publishers, 1965), esp. p. 39. On the basis of five-and-a-half years experience as a colleague of Kahn, this author can attest that Kahn was not confused concerning the relationship between the metaphor of his hypothetical escalation ladder and prospective reality. However, it is doubtful whether the same can be said for the majority of his readers.

42. See Spurgeon M. Keeny, Jr., and Wolfgang K. H. Panofsky, "MAD Versus NUTS," *Foreign Affairs*, vol. 60, no. 2, Winter 1981/82, pp. 287-304.

43. This author, a leading "NUT" according to Keeny and Panofsky (ibid.), agrees with radical critics of nuclear strategy that "endgame" analysis has tended to be distinguished by its rarity. There are excellent sociological, psychological, and bureaucratic reasons why this should be so, but it remains a distressing fact. This author must repeat a familiar refrain: If the United States is to have a theory of war worthy of the name, it must consider conflict as a unified continuum from the "war in peace" that exists today, through military engagement at all levels to the political circumstances of termination and settlement.

44. Karl von Clausewitz, *On War,* ed. and trans. Michael Howard and Peter Paret (Princeton: Princeton University Press, 1976), pp. 566-73.

45. Ibid., pp. 119–21.

46. This argument is central to Colin S. Gray and Keith B. Payne, "Nuclear Strategy: Is There a Future?" *Washington Quarterly,* vol. 6, no. 3, Summer 1983, pp. 55-66.

47. Edward N. Luttwak, "On the Meaning of Victory," *Washington Quarterly,* vol. 5, no. 4, Autumn 1982, pp. 17-26.

Chapter 4

NUCLEAR STRATEGY: THE RANGE OF CHOICE

There are cyclical trends, or oscillations, in the character and dominant strain of U.S. strategic thought. Following Harold and Margaret Sprout,[1] this author endorses the idea that the historically and culturally rich setting for U.S. defense policy provides a wide range of possibilities. American strategic culture is here viewed not as a constraint but as a tolerant license. The "American way of war" endorses both strategies of annihilation (*Vernichtungskrieg*) and attrition (*Ermattungskrieg*)—to resort to the enlightening distinction drawn by German military historian Hans Delbruck.[2] In the American Civil War, Sherman's march through Georgia and the Carolinas exemplified the former, while Grant's painful campaign before Richmond illustrated the latter. There is no real contrast. Grant's campaign of attrition permitted Sherman to wage a war of maneuver. In World War II, General Eisenhower pursued a strategy of attrition in a broad-front advance in Europe, for reasons both of Anglo-American accord (or tolerable discord) and logistic convenience,[3] while General MacArthur pursued the theme of maneuver-annihilation in his island-hopping campaign. MacArthur's Inchon landing during the Korean War was a further illustration of the U.S. capacity for pursuit of the maneuver-annihilation choice.

Although one may speak of American cultural proclivities with respect to strategy,[4] it is reassuring to appreciate that U.S. military experience points to a usefully wide range of policy options. Nonetheless, a close observer of the U.S. defense debate of the past ten years could not help but notice the doctrinal rigidity that has characterized different schools of thought. This is a somewhat bizarre phenomenon for a nation and culture that prides itself on its pragmatism. Americans should be the least likely to coalesce into doctrinally dogmatic, apparently exclusive groups.[5] Bernard Brodie wrote, accurately: "Strategic thinking, or 'theory,' if one prefers, is nothing if not pragmatic. Strategy is a 'how to do it' study, a guide to accomplishing something and doing it efficiently."[6]

It is fashionable to argue that a thousand flowers should be encouraged to bloom and that one person's theory is as good as the next. But this study prefers to hew closely to the ideal of U.S. pragmatism and to argue that many of the candidate nuclear postural/doctrinal concepts for the United States in the 1980s and 1990s have already been tested (short of battle, of course), have been found wanting, and should be identified clearly as inferior ideas.

In descending order of concern, the strategic nuclear forces of the United States are charged mainly with the deterrence of massive counterurban/industrial strikes; the deterrence of massive counterforce/counterpolitical strikes; and the ability to exercise coercive influence on behalf of forward-placed allies or exposed U.S. forces by way of extended deterrence. These tasks were outlined in detail by Herman Kahn in the early 1960s in his *On Thermonuclear War*[7] and *On Escalation*.[8] Nonetheless, the functions of strategic nuclear forces are ill-appreciated today.

The required character of a strategic force posture and the doctrine it expresses is (or should be) largely determined by the character of U.S. foreign policy that it is required to support and by the political-military capabilities and nature of likely adversaries. For a leading contemporary example, the official in-house and public debate over the MX ICBM and its basing mode has been conducted in a near vacuum with respect to foreign policy supportive duties, strategy, and Soviet strategic culture—all three of which are of central importance to the debate.[9]

The simplest task for the U.S. defense community is to design a strategic force posture capable of deterring a tolerably rational enemy from launching a massive attack against U.S. cities. Unfortunately the U.S. strategic nuclear defense planning for strategic nuclear forces cannot be restricted to such a task. Because the United States has global interests, its strategic forces must be relevant to the restoration of deterrence vis-à-vis some unfolding political-military catastrophe in a theater far from home. It is likely that it will be the United States which first feels moved to threaten and execute a central nuclear strike. This means that the question "Are we deterred?" should be asked first in Washington rather than in Moscow. This would be a reversal of the situation in the Cuban Missile Crisis of October 1962, when the Soviet Union had to decide whether to attempt to run the U.S. naval blockade.

One should respect the logical integrity of a policy-doctrinal opponent who marries a recommendation for a minimum U.S. strategic force posture (designed to assure the destruction of perhaps the one hundred largest Soviet urban areas) to a recommendation for the drastic retrenchment of U.S. foreign policy duties and interests. If the United States were to decide it had no vital foreign policy interests beyond the Western Hemisphere, a far less expansive definition of force-postural adequacy would be appropriate.[10] Also, were a policy-doctrinal opponent to recommend a general-purpose force posture (including tactical aviation, maritime assets, and theater-nuclear strength) that should, with the assistance of regional allies, be capable of defeating most local challenges, the deterrent and war-fighting burdens placed on the U.S. strategic nuclear force posture would be noticeably diminished.

A defense analyst is not totally at liberty to select a personally favored strategic posture and doctrine. The United States does have global commitments. The policy arguments with reference to the balance of power in Eurasia-Africa, vis-à-vis long-term U.S. security, that were persuasive in 1917 and 1940-41 are no less valid today. The United States could function minimally in isolation with the Soviet empire dominating the rest of the world outside the Americas, but that is not a world in which Americans should choose to live, and such an autarkic security condition of embattlement would have profound negative implications for the quality of U.S. domestic life.[11] On the defense postural side, although the United States could choose to stress general-purpose force capabilities, there are some enduring problems of geopolitics. The Soviet Union enjoys interior lines of communications vis-à-vis theater conflict around Eurasia if not Africa[12]; and scarcely less significant, the Eurasian allies of the United States have proved to be nervous of defense postural/doctrinal "improvements" that appear to make the international political system safer for local or theater wars.[13] In extreme circumstances, a favorable transformation of the capability of U.S. (and U.S.-Allied) nonnuclear forces may augment the incentives to nuclear employment on the Soviet side. As Richard Burt has argued, "Although emphasizing conventional forces will tend to raise the 'threshold' in local conflicts for the Western use of nuclear weapons, a conventional-emphasis strategy could actually provide the Soviet Union with incentives to escalate in time of war."[14]

Among the worst sins of policy-contending defense analysts is their inability to listen to arguments of the other side. This chapter offers a preferred policy option, but that option is offered on the basis of characterizations of alternatives that doctrinal opponents should acknowledge to be fair. All too frequently, policy debaters choose not to hear the arguments of the other side. This author has a clearly preferred strategic posture and strategic doctrine in mind, which is advanced here, but he is open to the logic of alternative perspectives. Opponents of the preferred option may cavil over the logic of the argument, but they should not be able to allege that their arguments have not been presented fairly.

Five nuclear-strategy options for the United States are discussed here in detail:

Option One: Mutual Assured Vulnerability

Option Two: Mutual Assured Vulnerability with Targeting Flexibility

Option Three: Counterforce and Countercontrol Preeminence with Recovery Denial

Option Four: Damage Limitation for Deterrence and Coercion

Option Five: Damage Limitation with Defense Dominance

Following President Carter's PD-59 and its associated nuclear weapons employment policy document (NUWEP-2) and NSDD-13 of President Reagan (October 1981—and Secretary of Defense Weinberger's NUWEP-82 of July 1982), the United States today is at Option Three in terms of declaratory policy.[15] This chapter recommends a move to Option Four, "damage limitation for deterrence and coercion." As demonstrated below, such a shift should affront relatively few current official shibboleths and is on the edge of being technically feasible in the late 1980s; it is certainly feasible for the 1990s. A preference for "damage limitation for deterrence and coercion" lies squarely in the center of U.S. strategic culture. It can be advanced as an essential today as the only concept that matches fully the foreign-policy supportive duties that continue to be placed on the strategic nuclear forces. Moreover, it should be technically and politically feasible in the calculable future.

Generally careful scholars have a tendency to debate cardboard adversaries.[16] It is often easier to debate preferred, largely fictitious doctrinal adversaries than real ones. In the following analysis of different postural and doctrinal ideas, there is no conscious tailoring of opposing arguments for the purpose of easy demonstration of error. The proponents of different beliefs concerning the U.S. strategic nuclear force posture may be judged to be in error, but their motives, patriotism, and so forth are not in question. No one has any hands-on knowledge concerning bilateral nuclear war, and virtually everything that is believed concerning what has or has not deterred has its basis largely in inferential, deductive reasoning.

Nuclear strategy, and more generally deterrence, is apt to be taught in a doctrinally permissive mode in universities and war colleges. Students are exposed to rival theorists and educated to believe that there is no source of authority on the subject. For example, at the Kennedy School of Government at Harvard, students have been asked to compare and contrast my opinions with those of Robert Jervis.[17] The idea that there may be a "correct" theory is inimical to contemporary liberal scholarship.

This chapter offers not merely a preferred posture and doctrine but also a posture and doctrine that the author claims to be logically correct, given U.S. foreign policy, the nature of the adversary, and what is technically feasible. Among the more debilitating features of the U.S. defense debate is the fact that truly fundamental doctrinal issues are seemingly never resolved. For example, the MX ICBM debate continues to be plagued by those who do not understand why the United States might require the services of the strategically distinctive characteristics of a land-based ICBM force.

Option One: Mutual Assured Vulnerability

The United States could decide that nuclear war-fighting and intrawar deterrent ideas were an illusion and that security could best be forwarded by advertising and acting programmatically on the basis of that decision. The matching U.S. strategic posture would be designed to hold at risk under all circumstances of attack a large number of Soviet urban areas and other economic targets believed essential to recovery from war. That number might be one hundred targets, or even more. U.S. strategic forces could be designed and sized for extravagant redundancy in that one might require each leg of the strategic triad to be independently capable of effecting the identified level of damage.

This posture and doctrine is often termed one of finite or minimum deterrence and is frequently accompanied by the opinion that even a handful of nuclear weapons on a handful of cities would deter and be viewed as a societal catastrophe.[18] However, the implementing posture identified is typically quite substantial. Since the late 1940s the United States has never had war plans that even approximated the idea discussed here.[19] Defense Department spokesmen in 1967 and 1968 often spoke and wrote in terms of the merits of mutual assured vulnerability, but critics (and admirers) of the idea of mutual assured vulnerability should not confuse rhetoric with operational policy.[20]

Nonetheless, the central core of reasoning which is the heart of mutual assured vulnerability arguments remains as significant in terms of public discussion of nuclear policy issues as it is insignificant, and of declining importance, in terms of recent defense planning. Although the defense intellectual and policy trend in the United States has been moving away from finite deterrence ideas, those ideas constitute an important and enduring landmark on the landscape of defense and continue to have a place in the schemes of advocates of other, more complex postures and doctrines.[21] Although there is no logically necessary connection between them, the idea of MAD (mutual assured destruction) underpins much of the reasoning in favor of the nuclear freeze. Many proponents of the freeze do not really care about the details of balance and imbalance or about the issue of what is a candidate to be frozen and what is not. In their view, nations are deterred by the general expectation that their principal urban centers would be devastated by nuclear weapons; plainly both superpowers assuredly would retain considerable overkill against that elementary target set.[22]

Rival debaters from contrasting schools of thought should be discouraged from debating caricatures of their opponents' arguments and public relations acronyms and pejorative slogans that are inaccurate. Nuclear strategy is a difficult enough subject when discussed fairly, without the added complication of deliberate or careless misrepresentation. Often the first victim of the oversimplified caricature is the author of that caricature himself. For example, advocates of a finite deterrence approach to the quality and size of the strategic force posture are interested in mutual vulnerability, not in executing mutual destruction. The politically effective acronym MAD (for mutual assured destruction) is not helpful for constructive debate. Similarly, proponents of a counterforce strategy with or without homeland defense do not constitute schools of nuclear *war-fighting* or nuclear-use theory.[23] This kind of pejorative shorthand does not facilitate genuine debate. Theorists of different doctrinal persuasions are arguing over theories of deterrence and over what could and should be done if deterrence fails.[24]

The discussion here and in subsequent sections of this chapter is organized by key elements in each strategy option, with commentary (generally of a critical kind) immediately following every statement.

1. Nuclear war would be a catastrophe unparalleled in world history.

Commentary
Nuclear war may or may not prove to be a catastrophe unparalleled in world history, but it is unlikely to be the functional equivalent of the cataclysmic biblical flood, notwithstanding the recent claim advanced by some scientists to the effect that nuclear war would probably trigger climatic changes that could be fatal to life on earth. The new apocalyptic vision is of the "nuclear winter."[25] In the thirteenth and fourteenth centuries, the Mongols and the bubonic plague were viewed in much the same eschatological terms in which many people today view nuclear war. Those "visitations from God" were terrible, but mankind remained in business. As Herman Kahn sought to establish more

than twenty years ago, catastrophe can come in different sizes and with very different consequences.[26]

2. Nuclear war could not be controlled or limited.

Commentary

Pessimists or realists may be correct in claiming, on the basis of no more evidence than the people they criticize, that nuclear war cannot be controlled or limited.[27] Any nuclear strategist who promises limited and controlled nuclear wars as the wars that would happen if his version of the requirements of deterrence were to break down should be distrusted. Mutual assured vulnerability theorists often do not appear to understand that their caricatured opponents are not promising nuclear wars of a guaranteed nature. Theorists who believe in intra-(nuclear) war deterrence are gambling that the Soviet High Command, in time of central nuclear war, will make decisions on the basis of calculations of self-interest functionally analogous to those of the United States. They could be wrong. Counterforce/damage limitation theorists do not exclude the possibility that catastrophe might occur should Soviet leadership either choose or be unable to cooperate. They argue that mutual assured vulnerability guarantees unlimited catastrophe, while their preference at least holds open the hope of containing the scale of potential damage.

3. Probably the greatest risk of nuclear war will stem not from Soviet leaders who are insufficiently deterred but from Western nuclear war-fighting theorists who may mislead policymakers into believing that nuclear weapons can be employed, like other weapons, as a political instrument.[28]

Commentary

Western theorists do not promise political advantage from controlled nuclear employment. They note that nuclear threats are integral to NATO strategy, that war could occur regardless of the quality of NATO's posture and doctrine, and that a theory of limited nuclear war is preferable to no such theory. Even the severely constrained nuclear campaigns envisaged by defense-minded Western analysts are acknowledged to be likely to entail a casualty rate so high as to give pause to, if not deter, any reasonable U.S. president.[29] No one is promising cheap nuclear wars or guarantees of very limited societal liability. Prudent commentators, regardless of doctrinal affiliation, are properly skeptical of the possibility of controlled nuclear war.

4. Most of the Western strategic literature that focuses on the need for credibility in deterrent threats and that worries about the alleged delicacy of the balance of terror[30] fails to understand how and why nuclear deterrence works. Nuclear deterrence "works" because sensible people, policymakers and people-in-the-street alike, are terrified by the prospect of nuclear war per se. No matter how large the escalatory leaps from a theater-conventional conflict to large-scale theater-nuclear war and then, most probably, to a large-scale central war,[31] any logical fragility in the credibility of the threat of such leaps is more than compensated for by the generalized fear of nuclear war. Unlike theorists of nuclear strategy, statesmen do not confuse the logic of the real world of political responsibility with the abstract, consequence-free logic of the strategic theorists' seminar room.[32]

Commentary

Western politicians are frightened by the prospect of nuclear war per se. Some U.S. strategists have not questioned this general truth; rather they have argued that Soviet governments have historically approached domestic human loss from a perspective different from that of their Western counterparts. In the 1920s and 1930s the Soviet governments killed close to 20 million of their own people, and the current leadership group was a party to and a survivor of that process.[33] While it is possible that Soviet leaders are deterred by the prospect of nuclear war, the available evidence on the Soviet Union may lead one to a different conclusion. While the Soviet Union does not want nuclear war and would not likely court the risk of its occurrence for positive gain, there is reason to believe that Soviet leaders would view nuclear war not as the end of history but as an experience to be survived and from which a resilient and objectively "progressive" society recovers. Moreover, it is likely that Soviet leaders fear nuclear war not so much for the amount of human and property damage it would cause but for the risk it would pose to Soviet political control at home. Sensitivity to human loss has not been a prominent feature of Soviet (or Russian) political culture. Anyone who believes that nuclear war should mean the same to Americans and to Great Russians should reflect deeply on the contrasting histories of the two societies.[34]

5. Nuclear weapons deter not only the employment of nuclear weapons by an enemy but also the kinds of actions that would or could create a political-military situation in which the use of such weapons would be judged to be much more likely. All people fear nuclear war, and roughly to the same degree.

Commentary

While it is true that all people fear nuclear war, it is not necessarily the case that all people fear nuclear war equally. Soviet military science teaches that nuclear missile weapons should be decisive in modern war and that, although a bilateral nuclear war will place unprecedented burdens on military organizations, it will also offer unprecedented opportunities for swift success.[35] Authoritative Soviet military opinion sees nuclear firepower in the context of long-range artillery.[36] Western analysts tend to err in their appreciation of the dynamics of an acute crisis. Soviet leaders are obliged by party doctrine to believe that their system will survive a nuclear war, and they also may well believe it because of the prudent provisions made over the past twenty years. Mutual assured vulnerability, insofar as it refers to the Soviet Union, is not a part of Soviet strategic culture. There is no persuasive evidence that the Soviet leadership is confident it can wage and win a central nuclear war. But it is correct to assert that Soviet leaders believe that victory is possible.[37]

6. Cultural nuance is not important in the nuclear deterrence system. A large nuclear war means the same thing to all cultures.[38]

There is some political-analytical merit in pointing to cultural distinctions between nations which may affect deterrent reasoning, but the sheer scale of damage that widespread nuclear war would impose renders discussion of operational nuclear strategy largely moot.

Commentary

It cannot be assumed that the level and kind of damage likely to be suffered in a central nuclear war are givens. Targeting "withholds" and other technical details such as yields selected, heights of burst, and so forth would be critically important to the scale of the catastrophe effected. Although nuclear weapons do come in inconveniently large packages of prompt energy release, careful weapon design, extreme accuracy, and concern for unwanted collateral damage can reduce potential societal damage by many orders of magnitude.[39] Soviet targeting style almost certainly does not lend itself to the idea of waging nuclear war in a severely constrained manner,[40] but the technical possibility cannot and should not be discounted. Finally, almost no matter how an enemy chooses to wage war, a nation can limit the damage that it could suffer.[41]

7. Nuclear war, should it occur, would hold the participants and many bystanding nations open to a limitless liability. Mutual assured vulnerability is not a posture and doctrine of choice. There is no choice. This posture and doctrine make a virtue of necessity. Strategists who insist on seeking out operationally interesting nuclear employment options in pursuit of an improved quality of prewar deterrence, intrawar deterrence, and/or damage limitation simply have not come to grips with the nature of nuclear war. There are no plausible theories offering a reasonable promise of bearable, survivable, recoverable, let alone winnable, nuclear conflict.[42]

Commentary

Nuclear war could prove to be a nonsurvivable, nonrecoverable catastrophe, but one can design war plans which should not lead to that dire result. No one is offering guarantees of nuclear war with strictly limited liability, but controlled and limited nuclear war is more likely to be a reality if it has been considered well ahead of time. The mutual assured vulnerability school of thought both discounts the Soviet evidence to the effect that Soviet society is unlikely to be in a condition even close to total vulnerability, and forecloses on the prospect of Western damage limitation in war. There would be an enormous difference between, say, 20 million and 120 million U.S. fatalities. Both are catastrophes, but the United States could recover from the former while it could not recover from the latter. This is far from a claim that 20 million fatalities would be acceptable.[43]

8. Nuclear weapons cannot be tamed. There cannot be a nuclear strategy with a human face. However, by choosing a nuclear arsenal that manifestly lacks the capability to threaten even a generous Soviet definition of its second-strike retaliatory force-level requirements, *and which itself is highly survivable,* the United States can diminish both crisis and arms race instabilities.[44]

Commentary

U.S. self-restraint in the region of strategic nuclear force deployment has had either no effect or an encouraging effect on Soviet defense planners. Soviet strategic force developments over the past decade have shown no evident sensitivity to U.S. crisis or arms race instability concerns. While some Western theories of stability show clearly, for example, that hard-target counterforce capability is destabilizing, Soviet weapon deployments do not betray any appreciation of this concern. In the Soviet view, prevention of war is totally a political function; it is the task of the armed forces to prepare efficiently for the actual conduct of war.[45] The idea that the detail of military posture could be important for political decisions in time of crisis, and perhaps even for military mobilization decisions relevant to the determination of war or peace, remains alien to the Soviet mind-set, at least insofar as it may have an impact on *Soviet* program decisions.[46]

The above characterization and commentary are offered as an intellectual anchor to one end of the policy-thought spectrum. They are not offered as representing accurately the contemporary beliefs of any particular individual or group of individuals.

Mutual assured vulnerability, though a vital part of U.S. thinking on defense even today, has never dominated official policy and planning. An important issue not discussed above is whether the U.S. defense community can enforce a mutuality of societal vulnerability. The Soviet Union has provided a plethora of hard bunkers for its political leadership cadre, blast shelters for its essential work force, and evacuation plans and fallout shelters for its general urban population. These plans may not work well in practice, but can a prudent Western defense analyst assume that they would fail catastrophically? More to the point, can a prudent Western defense analyst afford to assume that Soviet leaders would lack confidence in their preparations for war survival? It is not good enough to argue by assertion, as Glenn Buchan does, that "no decision-maker can have confidence that any preparations for war, in case deterrence fails, would be successful or that any recovery plans are realistic."[47]

Option Two: Mutual Assured Vulnerability with Targeting Flexibility

This second option corresponds roughly and generically to where most of the policy refugees from mutual assured vulnerability have evacuated intellectually. It is probably no exaggeration to claim that this second option is the thinking person's version of mutual assured vulnerability. There is no need to restate the arguments typically advanced in favor of mutual assured vulnerability. This policy option is particularly important because it represents an apparent way-station on the nuclear war-fighting course. Many people formerly associated with mutual assured vulnerability thinking will argue that they have never been opposed in principle to flexibility in strategic employment planning; that they have known since 1961 that SIOP planning provided several preplanned options, albeit large ones[48]; and that the conviction that mutual societal vulnerability is both a technological fact and desirable as a dampener of arms competitive urges carries no particular implications vis-à-vis the size and sequencing of targeting options. However, there are potential tensions between mutual assured vulnerability and flexibility in targeting. What follows are the essential characteristics of Option Two, with commentary appended.

1. Because mutual vulnerability is considered the ultimate basis for deterrence stability, neither nation should seek to acquire the means physically to limit damage to its homeland, through active and passive defenses or through development of offensive forces that threaten the survivability of the strategic retaliatory forces of the other side.

Commentary

Option Two suffers from the fundamental weakness of Option One. Should war occur and should the deterrent shock effect of initial, flexible strategic nuclear use not function as hoped, the United States could suffer a limitless catastrophe.[49] Moreover, the Soviet Union, although not eager to engage in nuclear combat, does not endorse the concept of *mutual* assured vulnerability.[50]

2. Because of the suicidal consequences of actually executing a major attack option against, for example, the Soviet "recovery economy," the credibility of such a threat is not high under most circumstances. Therefore, both to augment perception of a link between theater forces and strategic forces and to provide a president with employment options that might serve to restore deterrence in the course of a war without necessarily producing mutual holocaust, targeting flexibility is desirable.[51]

Commentary

Flexibility per se does not solve the U.S. president's self-deterrence problem. In principle, it should be more credible for him to threaten small- as opposed to large-scale nuclear strikes, but he would have to be profoundly fearful of the consequences of such action. Option Two does not contain a very plausible theory of escalation control, let alone dominance. If a political-military situation is sufficiently grave for a president to order very limited employment of central nuclear forces, it is reasonable to assume that both parties to the conflict have a truly major stake in the political outcome of the conflict. It is just possible that the shock of homeland-to-homeland nuclear use would restore deterrence, but it is hardly likely. It is more likely that the Soviet Union would respond by beginning to execute its central nuclear war plan. Should the Soviet Union, contrary to what the U.S. defense community thinks it understands about Soviet strategic culture, respond more or less in kind, what does the president try next? Among the more persuasive criticisms of limited nuclear options is the charge that they are unlikely to succeed in restoring deterrence.[52]

3. Flexibility should enhance deterrence, while its potential for damage in the realm of crisis and arms race stability can be minimized through endorsement of a posture that would be manifestly incompetent in fulfillment of preclusive counterforce missions. Moreover, the absence of BMD (ballistic missile defense), and of serious air defense and civil defense should reinforce declarations that strategic flexibility is not a move toward what is termed a war-fighting strategy.[53]

Commentary
Far from enhancing deterrence, Option Two could set in train a process of escalation that the United States could neither discipline nor win. A posture that because of determination not to enhance possible crisis and arms-race instabilities was obviously counterforce-incompetent would invite a counterforce-dedicated Soviet Union to escalate rapidly in search of victory or at least "useful advantage."[54]

4. Small nuclear strike options would be intended both to provide a deterrent shock and to carry the clear threat of "more to come, unless . . . ." By executing a limited nuclear option (LNO), one would have signaled determination through action in that two major thresholds would be crossed (use of central

nuclear forces, and employment most probably against the homeland of a superpower).[55] But the small scale and the nature of the attack would also signal unambiguously a willingness to exercise restraint and would constitute an invitation for the restraint to be reciprocated. In short, such limited employment would be part of a political bargaining process rather than constituting military action.[56]

Commentary

LNOs would more likely signal weakness than strength of will and capability. Given its quite well appreciated conflict style, the Soviet Union would probably be more impressed by what the United States did not do or was unable to do than by what actually was effected. Instead of signaling determination, very limited nuclear options would more likely be read in Moscow as signaling an extreme fear of nuclear war. Such fear is reasonable and sensible, but it is not the message that a U.S. president would want to transmit when he engages in what Thomas Schelling termed a competition in risk-taking.[57]

The addition of flexibility of strategic employment to a mutual assured vulnerability posture and doctrine seems more likely to produce defeat on the installment plan than effective intrawar deterrence. It would be imprudent to begin a small nuclear war unless one had on hand a capability for waging, surviving, and recovering from a large nuclear war.

Proponents of mutual assured vulnerability with flexibility have not been unmindful of the perils of the option, in terms of their core beliefs about stability and what deters.[58] The technical requirements for execution of limited nuclear options could drive one toward endorsing deployment of very accurate ICBMs which, so the argument goes, would be destabilizing because of their counterforce potential. Manned bombers and cruise missiles are inherently inappropriate for most LNO missions, because Soviet air defenses would not have been suppressed in advance. Also, SLBMs would be inappropriate because there may not be submarines on station to execute such missions; because communication may not be adequate, or even possible; and because an

SSBN comes with a distressingly large number of SLBM warheads (240 on the *Ohio* class) and betrays its position by launching even a single missile.[59]

With respect to the flexible and small-scale employment of nuclear weapons, extreme accuracy is desirable because the use of small warheads, creating the least possible collateral damage, is therefore feasible. In the future (in an era of "zero circular error probable," or perfect accuracy), it is possible that strategic nonnuclear weapons could accomplish many of the counterforce tasks that today either have to be allocated to nuclear-armed ICBMs or are not scheduled to be performed at all because of the problem of collateral damage.

In the immediate context of this discussion, flexibility implies small-scale employment. However, there is no inherent reason why flexibility need refer only to the very limited end of the employment spectrum. Flexibility is a strategy-neutral concept, long appreciated as a political and military desideratum. Soviet military science, too, endorses flexibility, though with specific referents that are far removed from the context of the U.S. debate of the mid 1970s.[60] Strategic flexibility at the level of principle is a difficult concept to oppose. It is no easy matter to argue for inflexibility. Critics of nuclear war-fighting strategies appreciate that endorsement of flexibility and the idea that central nuclear use might be controlled and limited places them dangerously close to accepting key elements of a "war-fighting" approach to nuclear planning. Agreement to some kinds of flexibility may open the floodgates to theories of controlled nuclear war, theories that adherents to Option Two do not want to endorse.

Option Two is dangerous in the eyes of adherents of classical mutual assured vulnerability doctrine because it might encourage the view (believed to be mistaken) that nuclear weapons can be used as political instruments and that nuclear war, in some dire circumstances, would be a sensible course to pursue and would probably remain limited.[61]

Option Three: Counterforce and Countercontrol Preeminence with Recovery Denial

Option Three is U.S. defense policy today. The U.S. defense community, coerced by the continuing adverse trend in the balance of forces, has addressed the vital strategic question "What do Soviet leaders find most deterring?" and has decided, almost certainly correctly, that the most fearsome threat in Soviet anticipation is the attenuation or loss of political control over the Soviet empire at home and abroad. In addition, it is well appreciated in the United States that a fully effective strike against the Soviet political control system is highly improbable, meaning that there are no easier options to the initial need to blunt Soviet military power directly.[62]

The most impressive defense-intellectual pyrotechnics of the nuclear era may have occurred in the "golden age" of 1955-65,[63] but the most valuable thought was probably registered from 1977 to 1979. The genuine appearance of the prospect of strategic inferiority, in the immediate context of the manifest failure of an erstwhile popular theory of strategic stability, stimulated the U.S. defense community to think through its evolving Soviet deterrence/war-fighting problem.[64] Unfortunately the intellectual task, in part for terms-of-reference reasons, was severely flawed with respect to its overall policy integrity. The principal flaw lay in the absence of a theory of damage limitation more robust than aspirations for a reciprocity in targeting restraint.[65] However, for the first time in the nuclear age the United States achieved a NUWEP design that reflected a sophisticated view of the distinctively Soviet adversary. The major characteristics of Option Three are described immediately below.

1. Option Three seeks to threaten the kind of damage that the Soviet Union should find most painful to suffer. The Soviet Union is deemed to fear not damage per se but damage of particular kinds. Today U.S. targeteers recognize that the relationships between state and society in the two superpowers are almost diametrically opposed. In the United States the state is, and ideologically is held to be, the servant of society, whereas the reverse is true in the Soviet Union. U.S. strategic targeting policy is reported to reflect this fact.

Commentary

To the extent that Soviet leaders can persuade Soviet citizens that the state's interest is really their interest, or that U.S. targeting policy cannot distinguish adequately between Soviet state and society, the fear and actual experience of war may serve to mobilize rather than fracture patriotic sentiment in the Soviet Union. The proposition that the Soviet leadership fears most for the continuity and effectiveness of its political tenure almost certainly is correct. But many unanswered questions remain with respect to the real vulnerability of the Soviet state to externally imposed shock. Even when they come with high accuracy and low yields, nuclear weapons inherently are weapons of mass destruction.[66] Much of the recent U.S. speculation about the possibility of forcing the regionalization of the Soviet Union is really fanciful.

2. The essential assets of the Soviet state must be held at nuclear risk. These assets are preeminently military in character. In short, the United States requires a strategic force posture that can inflict major damage on Soviet military power of all kinds. Above all else, there is a need to be able to threaten second-strike counterforce missions that would offset any benefit the Soviet Union might gain from a counterforce first strike (survivable U.S. strategic forces, in the second round of the war, would neutralize whatever gains Moscow had achieved in round one). Military power is the backbone and fundamental form of expression of the Soviet state, but it does not encompass the total areas of "essential assets." Because a war may be relatively long—say, up to six months—and because one should worry about the postwar balance, war-supporting industry also must be considered a prime target set.[67]

Commentary

The idea of the second-strike counterforce "equalizer" is attractive and has some theoretical merit,[68] but it appears to promise stalemate, which may or may not constitute a denial of Soviet victory. If the Soviet Union is faring well in a theater conflict, it is difficult to see why it should launch the first counterforce strike to which the United States needs to provide an offsetting reply. The most plausible reason for a massive Soviet counterforce blow against the United States would be a determination to achieve what John Erickson termed a " 'disruptive strike,' not unlike Soviet artillery practice in World War II."[69] This would be a case of the Soviet Union's striking first in the last resort in a central war, in preemptive anticipation of U.S. escalation. Plainly, the counterforce and countercontrol theme in U.S. nuclear strategy today is by no means confined in its relevance to the role of second-strike threat. In the words of the Scowcroft Commission,

> The Soviets must continue to believe what has been NATO's doctrine for three decades: that if we or our allies should be attacked—by massive conventional means or otherwise—the United States has the will and the means to defend with the full range of American power.[70]

3. The single most essential asset of the Soviet state, dependent though it is on the power of the Soviet armed forces, is the political control structure.[71] The United States should hold at prompt risk Soviet political leaders, the Central Committee *nomenklatura* at large,[72] the means of communication and command from Moscow to the provinces, and critically important elements of the KGB. If Soviet leaders know that their political system, as opposed to their society, is targeted reliably, they know that the United States has the capability to deny them victory.

Commentary

It is healthy for deterrence that the Soviet leadership be told that it and its means for enforcing domestic and imperial political control are targeted reliably. However, countercontrol targeting has some severe problems. The United States does not have a fully comprehensive understanding of the workings of the control structure in peacetime, let alone in wartime; it does not know what are and what are not essential targets; it does not wish to foreclose totally, early in a central war, on the possibility of negotiated war termination; and it does not have the ability to neutralize or blunt the Soviet retaliatory strike that should be expected to follow as a consequence of U.S. execution of a major countercontrol strike option. The U.S. defense community must consider whether a large countercontrol strike should be delivered up front as a bid for damage limitation, intended to destroy or paralyze the Soviet chain of political-military command, or whether the countercontrol strike option should be retained as the threat of last resort to protect U.S. cities.[73]

4. This strategy option has as its centerpiece a determination to deny victory to the Soviet Union, in Soviet terms.[74] The countervailing strategy of the last years of the Carter administration was dedicated to the mission of assuring effective deterrence through the promise of denying victory to the Soviet Union. The Soviet Union, so it was reasoned (and to a large extent is still reasoned), would not initiate a war that it was convinced it could not win with high assurance. Defense Secretary Weinberger has been explicit in explaining that his idea of denying the Soviet Union achievement of its war aims cannot rest on a threat of action against Soviet cities:

We disagree with those who hold that deterrence should be based on nuclear weapons designed to destroy cities rather than military targets. Deliberately designing weapons aimed at populations is neither necessary nor sufficient for deterrence. If we are forced to retaliate and can only respond by destroying population centers, we invite the destruction of our own population. Such a deterrent strategy is hardly likely to carry conviction as a deterrent, particularly as a deterrent to nuclear—let alone conventional—attack on an ally.[75]

Commentary

Denying victory to the Soviet Union is important, indeed essential. Unfortunately, a focus on victory denial is compatible with acquiescence in the prospect of Western defeat. In practice, a U.S. president venturing up the escalation ladder in or toward central nuclear war is likely to be much more interested in precluding a U.S. defeat than in denying victory to the Soviet Union. U.S. officials and strategic theorists have tended to commit the logical error of assuming that *the* problem is to determine what Soviet leaders find most deterring. That question is exceedingly important, but it is no more important than its logical strategic corollary: What deters a U.S. president? One should not design a U.S. strategic force posture and matching doctrine that cannot at least alleviate the self-deterrence problem. If it were the United States that felt moved to consider initiating a central nuclear war, for reason of impending theater defeat, the major deterrence problem would be American rather than Soviet.

5. Option Three rests on an awareness that, relative to the Soviet Union, U.S. strategic-force capability on most important measures of merit either has already slipped into the inferior category or soon will do so.[76] As with the U.S. Army's characterization of its operational problem in the mid 1970s, the issue here is how to fight outnumbered and win.[77] The idea central to former Defense Secretary Harold Brown's concept of the countervailing strategy was that of superior strategy. Even if the Soviet Union purchased a more impressive quantity of qualitatively not-too-dissimilar strategic capability, the U.S. defense community would design targeting plans for deterrence which, in the quality of fear they should produce, would offset the most serious U.S. strategic deficiencies. If, for example, U.S. strategic nuclear forces could hold at risk (at least in Soviet perspective) the most essential of Soviet state (as opposed to societal) assets, that should suffice to offset through its prospective denial of Soviet victory in war any advantage Soviet leaders might anticipate as a well-merited consequence of their newly acquired advantages in the gross figures of merit of strategic capability.[78] If all else fails, the United States will retain the ability to strike with devastating effectiveness against the Soviet recovery economy. Survival and a superior recovery potential are deemed vital by Soviet military science. In terms of putative deterrent effect, the United States would be able in extreme circumstances as its last resort to promise credibly to the Soviets that their ability to recover from World War III on a timetable likely to be politically acceptable could be fatally impaired.[79]

Commentary
Superior strategy is always desirable. It is acknowledged that strategic competition between the superpowers embraces a di-

mension of doctrinal rivalry.[80] The countervailing strategy is important historically in that it both recognizes the value of strategy and seeks to exploit distinctive Soviet vulnerabilities. However, the countervailing strategy, for all its genuine sensitivity to Soviet culture, neglected to consider a dominant reality of U.S. culture: A U.S. president could not intelligently hurt others if the certain consequence of such infliction of pain would be the delivery by the Soviet Union of a nuclear strike likely to inflict 100 million or more U.S. casualties. The United States could never effect a major attack option against the Soviet recovery economy—a somewhat elusive target set—because such action would result in a Soviet retaliatory strike that the United States could not survive. What is wrong with former Defense Secretary Brown's idea of a "countervailing strategy" is not that it promises to effect inappropriate kinds of damage on the Soviet Union but that it neglects the problem of American self-deterrence.

Contrary to the sense of strongly worded hostile Soviet commentaries, PD-59 did not provide guidance for a nuclear warfighter's manual. Reflecting accurately the two to three years of detailed research that preceded it, PD-59 adequately addressed the question of what prospective damage Soviet leaders fear most. Nonetheless, PD-59 had two major deficiencies—one internal, the other external.

First, it outlined a vision of U.S. counterforce and counterpolitical control activity in the SIOP which is the better part of ten years away from physical possibility (i.e., U.S. strategic forces cannot do the job in the 1980s). Second, as with its preceding strategic targeting review process, PD-59 ignored plausible connections between putative freedom of strategic offensive action and the ability to limit damage to the U.S. homeland.[81] Current nuclear policy is not so much wrong on strategic-logical grounds as it is incomplete. PD-59 was not a misstep, unless strategists choose to view it alone as the high-level doctrinal basis for strategic-force development over the years ahead.[82] Subsequent developments in policy guidance under the Reagan administration have recognized the importance of strategic defense.

By the end of the Carter administration, the small community of strategic targeting (*for deterrence*) aficionados was agreed that sub-SIOP level LNOs were relatively uninteresting[83]; that massive countereconomic (recovery) strike options were not useful either for "up front" declaratory-policy-for-deterrence purposes or for operational reasons; and that World War III could be either very short or relatively long (perhaps six months). There are grounds for entertaining some severe reservations over the plausibility of long World War III campaign scenarios.

Some defense analysts have recently become excited over the issue of the endurance of strategic forces and the National Command Authority, but somehow the idea that relatively long wars are possible moved from the status of idea to that of a driving planning assumption without obvious benefit of close strategic analytical scrutiny. On close inspection, one discovers that although six-month wars are possible, six-day or six-week wars are no less possible—and are indeed probably rather more plausible.[84] The problem here is that the U.S. defense community tends to be dominated by technicians and not by strategic thinkers. Strategic technicians typically have scant appreciation of likely operational issues, let alone of genuinely strategic considerations. Save in very exceptional cases, it is difficult to intrude strategic arguments into supposedly strategic policy decision processes. Typically, the major elements contending for preponderance are "technical fix" suggestions, considerations of managerial expediency, and the weight of vested bureaucratic interests in one, as opposed to an alternative, weapon system. Somehow the official U.S. defense community is not well organized either to generate or to respect truly strategic arguments.

It is frustrating to recognize that Option Three, where the United States is today, already has fractured the most important strategic cultural barrier; that is, Option Three recognizes the Soviet Union as a culturally distinctive adversary. The source of frustration lies in the appreciation that the United States, having broken free at the official level from strategic-cultural mirror-imaging, seems unable or unwilling to proceed logically the required additional mile to identification of an appropriate posture and doctrine. Having elected to take proper account of the uniquely Soviet aspects of the Soviet Union, U.S. officials are resisting the required further step of recognizing distinctively American problems.

A U.S. official, or extraofficial defense analyst, should know that his nation is and has always been acutely sensitive to American casualties. Save in the Civil War, high casualties have not been the U.S. military experience, and they have not been socially acceptable. Hence, the traditional American preference for profligate firepower to reduce U.S. casualty rates. Behind Option Three is full recognition that the Soviets have rejected the concept of a strategic stability reposing on the basis of mutual assured societal vulnerability. But Option Three, or PD-59 (plus the later refinements of NSDD-13), accepts the assured vulnerability of U.S. society. The Reagan administration has committed itself to exploration of the technical feasibility of strategic defense, and the president himself has certainly put the full weight of his office behind the vision of a United States no longer threatened by nuclear-armed ballistic missiles,[85] but that exploration and vision have yet to be translated into a damage limitation policy for the enhancement of deterrence.

Option Four: Damage Limitation for Deterrence and Coercion

Option Four, the strategy preferred in this study, is an evolution rather than a sharp break from the current U.S. posture and doctrine. It provides plausible answers to the more telling charges that can be leveled at Option Three. Option Four, in essence, would add a multifaceted homeland defense capability to the U.S. strategic posture. The title of this option was selected with care. The fundamental purpose of the strategic forces is to deter, or help deter, hostile acts against vital U.S. interests. The posture and doctrine outlined here should offer maximum discouragement to adventure and risk-taking on the Soviet part. However, deterrence is not sufficient as a statement of the mission of U.S. strategic forces. In addition to the negative task of dissuasion, those forces also have laid on them by foreign policy a range of possible "compellence" duties.[86] In other words, there may be occasions when the United States will have urgent political need to compel or coerce the Soviet Union to do things it is unwilling to do (e.g., to recall armies fighting successfully in the Persian Gulf area or in Western Europe). Such a coercive mission is compatible with a broad definition of deterrence. A limited exercise of strategic nuclear power could be effected for the purpose of "compelling" a Soviet withdrawal; in short, it would be intended to "restore deterrence."

Mischaracterization of Option Four as a war-fighting posture is politically damaging because the obvious elements of truth in the overall mischaracterization lend credibility to a mere caricature. It is true that damage limitation entails making preparation for the conduct of nuclear war, which is hardly a novel activity; the Soviet and U.S. defense communities have been making such preparations for more than thirty-five years. A proponent of damage-limiting strategy believes that the United States has no sensible choice other than to attempt to implement this idea in planning. In fact, this task is even worth doing badly.

Damage limitation is far from being a new idea. Prior to the nuclear age, armed forces provided damage limitation by serving as a hard shell around a society. To damage an enemy's society, one had first to defeat his army and navy; long-range aircraft, ballistic missiles, and nuclear weapons appeared to change the situation drastically. Henceforth, it was believed, intolerable damage could be inflicted, whether or not an enemy's armed forces were defeated in the field (the bomber and missile would always get through).[87]

In their doctrinal "revolution in military affairs" in the late 1950s, Soviet military theorists accommodated to the new technology by deciding that, far from overturning the existing wisdom of Soviet military thought, nuclear missile weapons would enable traditional tasks to be accomplished more swiftly and decisively. Neither then, nor since, did Soviet theorists accept fatalistically the proposition that nuclear weapons and their long-range means of delivery meant that the Soviet (and American) state and society must be totally vulnerable. The massive air defense program developed in the United States in the mid to late 1950s (though largely abandoned subsequently), in the context of serious official interest in passive civil defense and of evolution of a prospectively impressively competent counterforce capability, reflected endorsement of the common-sense logic of damage limitation.[88]

Damage limitation was doctrinally preeminent in the United States in the early years of Robert McNamara's tenure as secretary of defense. Year by year, however, it was relegated increasingly to a backstage role as Soviet strategic forces proliferated and were hardened and dispersed, and as the U.S. government embraced theories of crisis and arms race stability (on the presumption of the desirability of the mutual vulnerability of societies).[89]

Nonetheless, the idea of damage limitation persisted from the early 1960s to the present day in the form of a desire to see reciprocated restraint in targeting fuel the operation of an intrawar deterrence mechanism. For the past twenty years, the United States has been massively in the damage limitation business with respect to the scale of allocation of SIOP-assigned assets to counterforce missions. But since the mid 1960s there has been no expectation that truly effective damage limitation could be enforced through offensive counterforce action alone. Since active and passive defenses were eschewed for a mix of financial, technical, political, and strategic theoretical reasons, the hope for limitation of damage has reposed in belief in the possibility of reciprocated targeting restraint. One must deduce that even PD-59 of July 25, 1980, endorsed the theory of damage limitation through the functioning of an intrawar deterrence mechanism. The logical flaws in this theory, and the lack of operational prudence in its desired policy advice, lead this study to identify Option Four as a superior strategy. The characteristics of Option Four, plus a critical commentary, now follow.

1. In the absence of ability to hold down U.S. casualties and economic damage to a level "acceptable" in the context of the most important political interests at stake, U.S. strategy is either a bluff or heroically irresponsible.

Commentary

Casualties and economic damage in nuclear war cannot plausibly be held down to "acceptable" levels. In the words of Bernard

Brodie, "whether the survivors be many or few, in the midst of land scarred and ruined beyond all present comprehension, they should not be expected to show much concern for the further pursuit of political-military objectives."[90] Even if the United States were to endorse the multilayered damage limitation instruments favored by Option Four, it is probable that casualties would exceed the estimates of damage-limiting (for improved deterrence) proponents. Moreover, even if such proponents should be proved correct, how could a casualty list of 20 million be politically or morally "acceptable"?[91]

2. It is essential that the United States have an SIOP designed and selectively advertised publicly in general terms, so as to be able, in prospective execution, to promise denial of victory to Soviet leaders on their own terms.

Commentary
Denying victory is all very well, but considerable doubt remains over the real authority of the alleged Soviet official belief that victory is possible in nuclear wars.[92] Also, denying victory encourages the atavistic urge on the part of some U.S. strategists to press a "theory of U.S. victory" on the U.S. government.[93]

3. No matter how intelligent or clever U.S. strategic targeting design may be, the credibility of execution of such design is very low so long as the United States makes no noteworthy provision for the protection of its homeland against inevitable Soviet retaliation.

Commentary
The strategic logic of damage limitation is sound enough as strategic logic, but political decisions are not made totally in the light of abstract strategic logic. In practice, "our" leaders (and "theirs") would understandably be terrified of the possibility of nuclear war. Strategic analysis promising "only" 20 million casualties, even if believed (which it would not be), would not strengthen presidential resolve. A president would not need to be

told by the congressional Office of Technology Assessment that "the effects of a nuclear war that cannot be calculated are at least as important as those for which calculations are attempted."[94] Proponents of damage limitation through intrawar deterrence do not claim that such a deterrent mechanism will work, only that it might. They do not choose the possibility of limiting central war through targeting restraints, as opposed to limiting such a war through measures of damage limitation, because they do not believe the choice is real. Opponents of Option Four deny that a worthwhile measure of damage limitation is feasible, and they worry that the implementation of programs for ineffective damage limitation may mislead U.S. policymakers into believing that nuclear war can be waged and survived at "acceptable" cost.

4. If the sole problem were deterrence of a massive Soviet assault on North America, then the deterrence case for homeland damage limitation would be far less persuasive. Unfortunately, each of the more plausible, or less implausible, scenarios that involve possible employment of central nuclear systems must be structured (by reason of the probable geography of conflict) such that it is the United States that most needs to restore deterrence through the issuing of credible threats and, if need be, the implementation of nuclear strike plans.[95] The absence of protection for the U.S. homeland in these most likely circumstances should prove to have a paralyzing impact on the freedom of action of a desperate U.S. president.

Commentary

There is abstract merit in the argument that a United States that could limit damage to itself to a major degree should be able to extend deterrence on behalf of distant allies more credibly and reliably. However, in practice the allies of the United States would fear that the United States would develop a "margin of line mentality." Moreover, even if extended deterrence does require U.S. damage-limiting capability, the time is long overdue for NATO as a whole to diminish its security reliance on nuclear deterrence of all kinds—battlefield, theater, and strategic.[96] Extended deterrence was virtually a free gift, courtesy of the strategic-nuclear imbalance in favor of the United States which endured until the late 1960s. The challenge today should not be defined as ways in which that policy duty can be sustained in the face of a militarily far more capable Soviet opponent, but rather in terms of ways in which NATO Europe can diminish its debilitating client status vis-à-vis U.S. strategic forces.[97]

5. No one can predict the course of a central nuclear war. Intrawar deterrence may function as hoped. However, there is a significant chance that the superpowers would prove incapable of controlling a central war, while there is also a strong prospect that the Soviet Union would not be interested in any idea of control likely to prove tolerably congruent with U.S. wishes or interests. In short, for the extant official U.S. theory of strategic deterrence to function successfully in its hours of real test (acute crisis and war itself), an extraordinary degree of good fortune would have to bless its endeavors. Above all else perhaps, the Soviet war plan would in practice have to violate every known precept of Soviet strategic culture.[98] Moreover, for tacit and explicit negotiations the superpowers would have to be willing and able to communicate in the most physically, administratively, and psychologically stressful environment the world has ever known.[99] They might succeed, given extraordinary resilience and redundancy of equipment, historically unusual qualities of statecraft, and a great deal of luck. However, a prudent defense posture and doctrine can hardly be based on such an expectation.

Commentary

To cite the flaws in and problems with the prospective control and termination of a nuclear war, no matter how valid the citation, does not answer such questions as "Is damage limitation really feasible?" and "Would a damage-limiting strategy increase the risk of war occurring?" A good many of the opponents of damage limitation as a policy goal are driven not by the kind of strategic ideology of a particular definition of stability that John Newhouse celebrated in his account of the SALT I negotiations,[100] but by a deep skepticism that the United States can achieve a genuine defense against nuclear threats.

Ideologically, the Reagan national security bureaucracy is not opposed to Option Four, but in practice the shift from offense dominance to an offense-defense balance would strain any national security system. The U.S. problem today, as always, is strategic-conceptual at root. Defense decisionmakers wish to do the right things, but often they do not know what those things are. A secretary of defense may or may not be persuaded that a particular defense system would work adequately, at the level of a "technical fix," but he may not be open to argument on the subject of basic U.S. deterrent philosophy. The Reagan administration entered office with a mandate to correct the adverse trend in relative military preparation, but not with a mandate to adopt one or another alternative national security strategy.

The case for strategy is as strong in relation to the expanding Reagan defense budgets of the 1980s as it was in relation to the declining defense budgets of the 1970s. The Reagan administration is committed to rebuilding U.S. military power, but it also needs to rethink the conceptual basis of that power. The administration is willing to pour funds into new programs, but it may also have to be willing to engage long-standing domestic adversary constituencies in direct debate. The problem is one of political culture, for strategic culture and national style must reflect political culture.[101] Consideration of the strategy options discussed in this chapter should be informed by an awareness of the strategic arguments, but nations tend not to adopt defense postures and doctrines solely or even substantially for genuinely strategic reasons.

The kind of boldness shown by President Reagan in endorsement of ballistic missile defense on March 23, 1983, stands in contrast to the rhetorical caution displayed recently by the administration as a whole in expositions of its thinking on nuclear strategy. For reason of domestic and NATO-allied sensitivities—not to mention possible Soviet propaganda advantage—the administration has been anything but eager to expand public understanding of the essential elements of a war-fighting, or classical strategy, approach to nuclear deterrence. More troublesome even than the suspicion that policy declarations of a highly ambivalent kind may reflect or promote shifts in actual plans and programs is the emerging fact that the Reagan administration has permitted important aspects of its strategic nuclear policy to slide into a hostage status. In order to secure the congressional votes needed to sustain the MX ICBM program, the administration has entered into what may prove to be an imprecise Faustian pact with a powerful congressional arms-control lobby. Congress is too diverse in the views it contains to provide true leadership on strategic issues,[102] but it is more than capable of discouraging an administration from pressing policy ideas that are controversial—whether those ideas are controversial for good strategic reasons or not.

This is not to impugn anybody's motives. Rather, it is to suggest that a Congress determined to hold an administration's feet to the fire on the subject of arms control, and an administration recognizing that it can accomplish nothing of lasting value unless it can sustain a working consensus on its behalf in Congress, can between them produce some unsatisfactory strategic policy compromise outcomes.[103]

Persuasive arguments can be designed pro and con on a wide range of strategic postures and doctrines. No one's theory of intra-(nuclear) war deterrence or damage limitation has yet been road-tested, while the possible reasons that war is prevented are so various, and impossible to assay, that one cannot sensibly point to the evidence of nuclear peace since 1945 as clear and unambiguous proof of the merit of any particular theory of deterrence. Option Four is designed to cope with an unusually stressful set of circumstances, where deterrence is particularly difficult to effect or simply does not apply. U.S. diplomacy day-by-day certainly does not need the support of Option Four, but in the event of the true "war is in sight" crisis, nothing less than Option Four would likely be adequate, while even this option may not suffice to deter or to hold wartime damage down to an "acceptable" level.

Option Five: Damage Limitation with Defense Dominance

The reasoning behind Option Five tends to be more prudential and technical than strategic.[104] As Donald Brennan wrote in 1969:

I do not believe that any of the critics of BMD have even the beginnings of a plausible program for achieving major disarmament of the offensive forces by, say, 1980. Many of them seem committed to support a strategic posture that appears to favor dead Russians over live Americans. I believe that this choice is just as bizarre as it appears; we should rather prefer live Americans to dead Russians, and we should not choose deliberately to live forever under a nuclear sword of Damocles.[105]

From the mid 1960s until the early 1980s, the United States endorsed a theory of strategic stability which held that the active and passive defense of a superpower homeland is not merely infeasible but undesirable as well. Nuclear peace was judged to rest most reliably on the accurate perception by all policymakers and policy-relevant publics that, in the event of war, catastrophe could and probably would be unlimited. In keeping with this U.S. belief, SALT I and the abortive SALT II licensed an offense-dominant and unchallenged strategic nuclear environment.

It is not implied here that dominance for the offense was chosen solely because of a particular strategic ideology (deterrence through mutual assured vulnerability), but it is claimed that ideology played a major role.[106] Proponents of offense-dominance have political, technical, and ethical, as well as strategic-theoretical, cases to advance. The discussion that follows summarizes major elements in the case for Option Five, Damage Limitation with Defense Dominance, with appropriate commentary.

1. Defense is possible. A defense-dominant world might be enforced through unilateral U.S. changes in posture, but it would be accomplished more readily if strategic offensive arsenals were reduced drastically by formal arms control agreements.[107] The world envisaged here would be defense-dominant, not defense-exclusive. Perfect defenses are not possible, but even imperfect defenses would save many lives. In addition, strategic defenses, especially if they comprised a combination of different kinds of systems, would have to add greatly to the uncertainties that must promote deterrence-enhancing doubts in the minds of attack planners.

Commentary

It is not obvious that a defense-dominant world, as opposed to a useful level of defense for damage limitation, is technically feasible. Even if it is feasible, the pertinent time horizon for full operational capability probably is on the order of twenty to thirty years.[108] Furthermore, although the Soviet Union is known to favor assured survival, it is not known to favor mutual assured survival. There is no good reason to suppose that the Soviet Union would choose to cooperate in effecting an orderly transition from an offense-dominant world to a defense-dominant world. An imperfect defense in the nuclear age is no defense at all—even worse, it is the illusion of a defense that could be dangerous if political leaders place undue confidence in it. So many and so troublesome are the uncertainties that beset planners and policymakers with regard to the prospective operational effectiveness of offensive forces alone that it is far from self-evident that the additional uncertainties posed by the presence of active defenses are needed.[109]

2. There is nothing inevitable about the occurrence of major war, but July-August 1914 did happen. An international order enforced in part by latent and irregularly explicit nuclear threats is a world that one day will see a nuclear war. Good management and good luck have seen us to the early 1980s without a nuclear war, but the problem is the prevention of nuclear war *forever*. In an offense-dominant strategic world, it would take only one major failure (or possibly even minor failure) of the deterrence system for the United States to be out of business permanently. It is worth noting the judgment that the policymakers of the summer of 1914 were not noticeably less competent individuals than the policymakers of today. It is intolerable that Western and Soviet civilization should forever be totally vulnerable to a single sequence of major crisis mismanagement.

Commentary

Of course a breakdown could occur in the existing system of nuclear threat; that is not at issue. However, none of the major players in the 1914 case tried very hard to avoid war, a negligence that is hardly likely to be repeated in an era of nuclear risk. What is at issue is whether such a breakdown would or would not be more likely to occur should the major powers effect the postulated transition to defense-dominant strategic arsenals. It may be accurate to paint a glowing picture of a world freed from dread of the nuclear sword of Damocles, as President Reagan did on March 23, 1983, but such a world would by definition offer vastly reduced painful consequences for folly, adventure, and mismanagement. Such a reduction could have a marked negative impact on the incidence of major crises and wars.

3. Nuclear deterrence would not cease to function in a defense-dominant world. Active defenses, though impressive, would not be totally leakproof, and in the minds of politicians residual doubts would remain over just how efficient their untested defenses would prove to be in combat. But those same doubts would have to be harbored on a magnified scale in the minds of policymakers in a government considering the risks of attack.

Commentary
It is all very well to argue that nuclear deterrence would still function in a defense-dominant world, but such a major reorientation in posture and doctrine could be accomplished only by means of strongly phrased arguments to the effect that society would no longer fatally be at risk to the consequences of deterrence breakdown. If Soviet military theorists, today in an offense-dominant world, can argue that "victory" (embracing survival and recovery) is possible, how would the prospects of "victory" appear in a world where the balance of strategic armaments deliberately had been tipped massively in favor of defense? It seems as if proponents of defense dominance want to have it both ways—they argue that strategic defenses would strengthen deterrence while in the same breath suggesting that a defended world is a world that has transcended the perils of the nuclear deterrence system. Ben Adams, for example, claims: "Strategic defenses not only contribute to deterrence, they can repulse an attack, shield the homeland, protect the offense, and provide an opportunity to

regain the initiative, rather than conceding it to an enemy *as a deterrent strategy does.*" (Emphasis added.)[110]

In theory there need be no opposition between defense and deterrence. As Glenn Snyder made clear more than twenty years ago, one can deter by denial and/or punishment.[111] Proponents of defense often seem to be needlessly ambivalent in their attitude toward deterrence.

4. Assuming a bilateral superpower move to a defense-dominant posture, the quantity and quality of societal damage that could be imposed on the Soviet Union would be reduced to the point where the United States and its friends and allies would worry about the feasibility of the extended deterrent duties that traditionally had been charged to the strategic offensive forces.[112] The argument probably would proceed as follows: The credibility of U.S. offensive action would be high because of the limited liability to which U.S. society could be held in a defense-dominant world, but the quality of nuclear deterrence would be reduced by virtue of the same condition. (In effect, this world is "safer" than the present one for war, nuclear and otherwise.) Aside from the residual uncertainty about the real, operational effectiveness of heavily deployed active defenses, the Soviet Union should be compelled by U.S. and NATO-European (and Chinese) conventional and battlefield nuclear deployment to doubt its ability to prevail in the theater. While extended deterrence prospectively would be both politically more credible and strategically less effective, its necessity would and should be diminished by a NATO prepared to conduct so effective a local nonnuclear defense that the burden of decision to escalate or not to escalate would rest on Soviet shoulders.[113]

Commentary

Proponents of damage limitation through dominant defenses cannot cope plausibly with skepticism about the feasibility of extended deterrence. A residual fear of nuclear war would exist in a supposedly defense-dominant world, but the risk calculus of a potential aggressor would have to be affected negatively (for stability) as the aggressor considered the range of possible painful consequences of his actions. It is an exaggeration to claim that strategic defense dominance would make the world safe for theater conflict, from the perspective of the superpowers, but that exaggeration does point to a certain crisis in U.S.-allied relations. The transition to a defense-dominated U.S. strategic nuclear posture would require a near-revolution in NATO capability, if not an agreed NATO strategic doctrine. MC (Military Committee) 14/3 of 1967, with its endorsement of the concept of flexible response, could be retained as a general framework, but the balance of deterrent effect would have to shift down the scale of potential violence in favor of conventional forces.[114]

5. Proponents of Option Five should grant that they have a serious problem in the area of extended deterrence, or with respect to the feasibility of effecting a conventional-force partial substitute for extended deterrence, but they argue that that problem needs to be set against the prospect of removing the danger of unlimited nuclear catastrophe from the human race.

Commentary

It is highly improbable that friends and allies of the United States would be willing to consider altruistically the possible merits of Option Five for the human race or, as Jonathan Schell put it, "the fate of the earth,"[115] given that this option could have potentially catastrophic implications for their own national security. It would be argued by NATO-European officials and commentators that the United States foolishly was choosing to destroy the structure of mutual nuclear deterrence which helps to preserve peace. For fear of the possible ultimate consequences for American society, the United States would be designing an international military order wherein its allies would be expendable. European reactions to President Reagan's speech of March 23, 1983, were overwhelmingly negative.[116] European officials generally were less impressed with the idea that a defended United States might be a more reliable ally than they were with the fact that superpower strategic defenses would challenge the viability of third-party nuclear missile deterrents. Also, Europeans were, and remain, anxious lest a strong U.S. bid for strategic defenses lead to the demise of the ABM treaty. In Europe little comfort was drawn from the technical fact that space-based BMD systems could defend Western Europe as well as the United States. As a near visceral level, some Europeans have seen in the president's proposal additional evidence of a U.S. proclivity to seek security in technology rather than in political engagement.[117]

This author is sympathetic to the motives of those who favor Option Five, a defense-dominant strategy, and has an open mind about the technical feasibility and strategic desirability of the idea. For example, it is possible that the current research on directed energy weapons will produce useful though not leakproof defense against ballistic missiles, while defense against cruise missiles and penetrating manned bombers similarly could be successful. But such a prospect would serve to direct military research heavily both into the region of countermeasures and toward development of offensive weapons impervious to such defenses.[118] To cite the certainty that each superpower would seek to neutralize the effectiveness of the defenses of the other is not to cast a vote against such defenses. The United States builds tanks and aircraft, even though effective antitank and antiaircraft weapons exist.

Option Five was advocated strongly by Donald Brennan in the late 1960s,[119] but since then it has not attracted a politically significant following. Whether or not the Reagan administration is able to begin the process of transforming the president's vision of effective BMD into a real. sustained drive for homeland defense remains to be seen.[120] Option Five is compelling, notwithstanding the moderately skeptical commentary provided above. It really is intolerable, and foolish, to *choose* to live under a nuclear sword of Damocles indefinitely. One day that sword is likely to fall. However, if advocates of a defense-dominant strategy in the 1980s are to command the respectful hearing that their idea merits, they must show they are aware of and sensitive to the major strategic, technical, and political objections to their preference.

There may seem to be a theoretical artificiality about the discussion of five discrete strategy options. After all, the U.S. government makes its strategic program decisions incrementally. There never seems to be a right time for shifting gears to a different strategy. Each administration is the legatee of the program decisions of its predecessors and inherits weapon programs, and strategy and targeting plans, that cannot feasibly be terminated or substantially reoriented in the near term because of large sunk costs in hardware that would not be well suited to a new vision of an adequate strategy. Notwithstanding these considerations, the case for strategy reappraisal is overwhelming. The American public has a right to expect that defense policy should be treated as an economic problem as well, that is, as a problem in the efficient allocation of scarce resources. It is not politically acceptable to testify before Congress simply to the effect that more effort is required across the board.

Individual weapon programs should make strategic sense within the framework of a coherent theory of war, for improved deterrent effect and as insurance should war occur nonetheless. The lead time on weapon development poses serious problems for senior officials attempting to compose a coherent military posture, but at the least those officials should have a clear vision of whither they are intending, and why.

NOTES

1. Harold and Margaret Sprout, *The Ecological Perspective on Human Affairs with Special Reference to International Politics* (Princeton: Princeton University Press, 1965).

2. Cited in Walter E. Kaegi, "The Crisis in Military Historiography," *Armed Forces and Society,* vol. 7, no. 2, Winter 1980, p. 311.

3. See Russell F. Weigley, *Eisenhower's Lieutenants: The Campaign of France and Germany, 1944-45* (Bloomington: Indiana University Press, 1981); and Martin van Creveld, *Supplying War: Logistics from Wallenstein to Patton* (Cambridge: Cambridge University Press, 1977), chap. 7.

4. See Russell F. Weigley, *The American Way of War* (New York: Macmillan Co., 1973); and Colin S. Gray, "National Style in Strategy: The American Example," *International Security,* vol. 6, no. 2, Fall 1981, pp. 21-47.

5. This phenomenon is noted in John E. Mack, "Psychosocial Effects of the Nuclear Arms Race," *Bulletin of the Atomic Scientists,* vol. 37, no. 4, April 1981, esp. pp. 20-22; and in Theodore Ropp, "Strategic Thinking Since 1945," in Robert O'Neill and D. M. Horner, eds., *New Directions in Strategic Thinking* (London: Allen & Unwin, 1981), p. 5.

6. Bernard Brodie, *War and Politics* (New York: Macmillan Co., 1973), p. 452.

7. Herman Kahn, *On Thermonuclear War* (Princeton: Princeton University Press, 1969).

8. Herman Kahn, *On Escalation: Metaphors and Scenarios* (New York: Praeger Publishers, 1965).

9. This example is important because the MX ICBM epitomizes the kind of strategic capability the United States needs because of its extended deterrent duties (largely in Europe).

10. Except for the consideration that a Soviet Union unconstrained by U.S. power in Eurasia-Africa should come to pose a far more substantial threat to the United States than is the case even today, by virtue of Soviet ability to mobilize the economic assets of Eurasia on its own behalf. For detailed analyses providing useful background for those judgments, see James T. Lowe, *Geopolitics and War: Mackinder's Philosophy of Power* (Washington: University Press of America, 1981); and Colin S. Gray, *Basic U.S. Choices, 1982-2000* (Fairfax, Va.: National Institute for Public Policy, March 1983).

11. See Donald G. Brennan, *National Security in Fortress America,* HI-1974-P (Croton-on-Hudson, N.Y.: Hudson Institute, May 2, 1974). The social-political cost of true isolation tends to be overstated by U.S. internationalists/interventionists.

12. But see Albert J. Wohlstetter, "Illusions of Distance," *Foreign Affairs,* vol. 46, no. 2, January 1968, pp. 242-55.

13. The Reagan administration has rediscovered what Robert McNamara learned in the early 1960s—that forward-placed potential "battlefield" countries are strongly resistant to that prospective status. An important though indeterminate fraction of the hostile NATO-European official reaction to President Reagan's endorsement of the concept of homeland defense on March 23, 1983, flowed from anxiety lest a defended America in the future be far more relaxed over the dangers of "theater" war than is the case at present.

14. Richard Burt, "Reassessing the Strategic Balance," *International Security*, vol. 5, no. 1, Summer 1980, pp. 49-50.

15. This ignores the uncertain implications for "policy" of President Reagan's recent endorsement of the concept of homeland defense. See "President's Speech on Military Spending and a New Defense," *New York Times*, March 24, 1983, p. 20. See, also, Jeffrey Richelson, "PD-59, NSDD-13 and the Reagan Strategic Modernization Program," *The Journal of Strategic Studies*, vol. 6, no. 2, June 1983, pp. 125-46; and Desmond Ball, *Targeting for Strategic Deterrence*, esp. pp. 20-23.

16. This point is central to Glenn C. Buchan, "The Anti-MAD Mythology," *Bulletin of the Atomic Scientists*, vol. 37, no. 4, April 1980, pp. 13-17.

17. To be specific, students were asked to compare and contrast Colin Gray, "Nuclear Strategy: The Case for a Theory of Victory," *International Security*, vol. 4, no. 1, Summer 1979, pp. 54-87, and Robert Jervis, "Why Nuclear Superiority Doesn't Matter," *Political Science Quarterly*, vol. 94, no. 4, Winter 1978-80, pp. 617-33.

18. See McGeorge Bundy, "To Cap the Volcano," *Foreign Affairs*, vol. 48, no. 1, October 1969, p. 10.

19. The character and scale of damage deemed feasible in 1946-48 were driven by severe contemporary stockpile limitations. See David Alan Rosenberg, "U.S. Nuclear Stockpile, 1945 to 1950," *Bulletin of the Atomic Scientists*, vol. 38, no. 5, May 1982, pp. 25-30; and idem, "The Origins of Overkill: Nuclear Weapons and American Strategy, 1945-1960," *International Security*, vol. 7, no. 4, Spring 1983, pp. 3-71.

20. See Henry S. Rowen, "The Evolution of Strategic Nuclear Doctrine," in Laurence Martin, ed., *Strategic Thought in the Nuclear Age* (Baltimore: Johns Hopkins University Press, 1979), pp. 131-56.

21. Glenn Buchan is correct in asserting a distinction between finite deterrence and MAD in that the former implies a relatively modest scale of nuclear weapon deployment while the latter need not. However, like it or not, MAD is a theory of finite deterrence. See Buchan, "The Anti-MAD Mythology," esp. p. 15.

22. For a full-scale critique of the nuclear freeze and its associated ideas, see Colin S. Gray and Keith B. Payne, *Nuclear Freeze? Implications for American Security* (Cambridge, Mass.: Abt Books, 1984).

23. Some liberal theorists, enraged by the MAD acronym, coined the rival acronym, NUT (for nuclear utilization theory). See Buchan, "The Anti-MAD Mythology," p. 13n.

24. Even Robert Scheer concedes that "they [neo-hawks] are not eager for nuclear war anymore than I am" (*With Enough Shovels: Reagan, Bush, and Nuclear War* [New York: Random House, 1982], p. 121). However, he proceeds to assert that "they are as eager for confrontation as they are opposed to accommodation with the Soviet Union" (p. 121). As one of the people identified by Scheer as a neo-hawk, I have some difficulty recognizing my views in his assertion—"eager for confrontation"?

25. A scientific conference held in Washington in the fall of 1983 on "The World After Nuclear War" was the occasion for the presentation of the "nuclear winter" thesis. For the argument that it is the possibility of the fatal destruction of the ecosphere that should drive our approach to problems of security, see Jonathan Schell, *The Fate of the Earth* (New York: Alfred A. Knopf, 1982).

26. Kahn, *On Thermonuclear War,* esp. chap. 2.

27. See Desmond Ball, *Can Nuclear War Be Controlled?* Adelphi Paper no. 169 (London: IISS, Autumn 1981); and John D. Steinbruner, "National Security and the Concept of Strategic Stability," *Journal of Conflict Resolution,* vol. 22, no. 3, September 1978, pp. 411-28.

28. See Leon Wieseltier, *Nuclear War, Nuclear Peace* (New York: Holt, Rinehart and Winston, 1983), esp. chap. 2. Wieseltier casts me as a major villain in what he terms "the Sovietization of American strategy" (pp. 47-53).

29. See Colin S. Gray and Keith B. Payne, "Victory Is Possible," *Foreign Policy,* no. 39, Summer 1980, esp. pp. 25-26.

30. The classic statement remains Albert J. Wohlstetter, "The Delicate Balance of Terror," *Foreign Affairs,* vol. 37, no. 2, January 1959, pp. 211-34.

31. For an exceptionally clear statement of the structure of NATO's deterrence reasoning, see General Bernard W. Rogers, "Greater Flexibility for NATO's Flexible Response," *Strategic Review,* vol. 11, No. 2, Spring 1983, pp. 11-19.

32. See Bundy, "To Cap the Volcano," pp. 10-11, 13.

33. Some critics of U.S. so-called war-fighting theorists have come to believe that "war-fighters" are arguing that because the Soviet Union suffered 20 million casualties in the Great Patriotic War it would be willing to take

casualties on a comparable scale again. Judiciously framed, the argument has the following elements: The Soviet state has a patrimonial view of its human and other economic "property"; it is indifferent to individual human suffering or loss; it knows that the loss of 20 to 40 million people, in addition to vast economic devastation and dislocation, is "survivable" in "state entity" terms. One should not extrapolate Soviet human losses from 1917 to 1945 into a putative nuclear war case in a simpleminded fashion. However, it is a difficult exercise to seek to deny the qualitative difference between U.S. and Soviet views of human loss. For a good recent example of a poorly framed argument in this area, see Buchan, "The Anti-MAD Mythology," p. 17. All American students of the Soviet-American deterrence relationship should read Karl A. Wittfogel, *Oriental Despotism: A Comparative Study of Total Power* (1957; reprint ed., New York: Vintage Books, 1981).

34. Great Britain's mercenary Hessian soldiery of the 1770s may not have behaved like Boy Scouts, but the character and extent of their wickedness were light-years removed from the Russian experience of Mongol conquest and the subsequent "Tartar yoke."

35. See Joseph D. Douglass, Jr., and Amoretta M. Hoeber, *Soviet Strategy for Nuclear War* (Stanford, Calif.: Hoover Institution Press, 1979), chap. 2.

36. This is scarcely surprising, given that the Soviet ballistic missile programs in the 1940s and 1950s delivered operational weapons to the artillery reserves of the Supreme High Command. The first commander in chief of the Strategic Rocket Troops was Marshal of Artillery M. I. Nedelin. See Harriet F. Scott and William F. Scott, *The Armed Forces of the U.S.S.R.* (Boulder, Colo.: Westview Press, 1979), pp. 133-41. Also see John Erickson, "The Soviet Military System: Doctrine, Technology, and 'Style,' " in John Erickson and E. J. Feuchtwanger, eds., *Soviet Military Power and Performance* (Hamden, Conn.: Shoe String Press, pp. 23-31.

37. See Robert Bathurst, "Two Languages of War," in Derek Leebaert, ed., *Soviet Military Thinking* (London: Allen & Unwin, 1981), p. 32; and Henry Trofimenko, *Changing Attitudes Toward Deterrence,* ACIS Working Paper no. 25 (Los Angeles: Center for International and Strategic Affairs, UCLA, July 1980), pp. 23-25. The quality of recent Western scholarship on Soviet military thinking is indicated in John Erickson, "The Soviet View of Deterrence: A General Survey," *Survival,* vol. 24, no. 6, November/December 1982, pp. 242-51; William G. Hyland, "The U.S.S.R. and Nuclear War," in Barry M. Blechman, ed., *Rethinking the U.S. Strategic Posture* (Cambridge, Mass.: Ballinger Publishing Co., 1982), chap. 3; March E. Miller, *Soviet Strategic Power and Doctrine: The Quest for Superiority* (Washington: Advanced International Studies Institute [in association with the

University of Miami], 1982); Nathan Leites, *Soviet Style in War* (New York: Crane, Russak, 1982); David Holloway, *The Soviet Union and the Arms Race* (New Haven: Yale University Press, 1983); and P. H. Vigor, *Soviet Blitzkrieg Theory* (New York: St. Martin's Press, 1983). The most recent unquestionably authoritative statement of Soviet military thinking is Marshal of the Soviet Union Nikolay Vasilyevich Ogarkov, *Always in Readiness to Defend the Homeland,* trans. JPRS L/10412, March 25, 1982 (Moscow: Voyenizdat, January 1982).

38. See Buchan, "The Anti-MAD Mythology," p. 17.

39. This is not to claim that nuclear war *will* be survivable, only that it is more likely than not that it will be, particularly if the United States invests in multilayered protection of U.S. society. This argument should have been settled by Kahn's *On Thermonuclear War* more than twenty years ago. It is not helpful or sensible to argue that "the burden of proof must be on any decision-maker who would consider a general nuclear war to convince himself . . . that the risks of a nuclear war are manageable, and that the outcome could be predicted with sufficient certainty to make war a viable policy option" (Buchan, "The Anti-MAD Mythology," p. 14). Nuclear war is important as a set of options in U.S. defense planning and must be a viable policy option. Appearances to the contrary notwithstanding, this is not to confuse a logical necessity of policy with attainable reality. Whatever Donald Hanson may believe, this author can distinguish between verbal formulas and operational practice (see Donald W. Hanson, "Is Soviet Strategic Doctrine Superior?" *International Security,* vol. 7, no. 3, Winter 1982, esp. pp. 72-73). Critics of counterforce, damage-limiting strategies often give the appearance of believing they are saying something original and significant when they cite the horrors of nuclear war. Nuclear strategists do not need to be reminded of the penalty society could pay should deterrence fail. See Colin S. Gray, "Nuclear Strategy: A Regrettable Necessity," *SAIS Review,* vol. 3, no. 1, Winter-Spring 1983, pp. 13-28.

40. See Trofimenko, *Changing Attitudes Toward Deterrence,* p. 25; and Leon Gouré and Michael J. Deane, "The Soviet Strategic View," *Strategic Review,* vol. 8, no. 4, Fall 1980, pp. 79-84.

41. This certainly is the Soviet view. See Jacqueline K. Davis et al., *The Soviet Union and Ballistic Missile Defense,* Special Report (Cambridge, Mass.: Institute for Foreign Policy Analysis, 1979); and Michael J. Deane, *Strategic Defense in Soviet Strategy* (Washington: Advanced International Studies Institute [in association with the University of Miami], 1980).

42. This is a major conclusion in Lawrence Freedman, *The Evolution of Nuclear Strategy* (London: Macmillan & Co., 1981); see esp. chap. 26.

43. Robert Scheer has claimed falsely that this author (in "Victory Is Possible") has "specified that 20 million U.S. fatalities would represent an acceptable cost in a nuclear war." See Scheer, "Pentagon Plan Aims at Victory in Nuclear War," *Los Angeles Times,* August 15, 1982, p. 1. See the discussion in Gray, "Nuclear Strategy: A Regrettable Necessity," pp. 15, 21. Secretary of Defense Weinberger flatly denied the accuracy of Scheer's reporting in U.S. Congress, Senate, Committee on Foreign Relations, *U.S. Strategic Doctrine, Hearing,* 97th Cong., 2d sess., December 14, 1982, pp. 24-25.

44. In its endorsement of military effectiveness as a requirement for effective deterrence, the Scowcroft Commission was sensitive to the instability arguments mentioned here in the text. In its report the commission states that the deployment of one hundred MX missiles "would provide a means of controlled limited attack on hardened targets but not a sufficient number of warheads to be able to attack all hardened Soviet ICBMs, much less all of the many command posts and other hardened military targets in the Soviet Union" (*Report of the President's Commission on Strategic Forces,* April 1983, p. 18; hereafter cited as Scowcroft Commission Report). For a strong statement of the view that one hundred MX missiles provide a nowhere near adequate quantity of prompt counterforce forepower to offset Soviet SS-17s, SS-18s, and SS-19s, see A. G. B. Metcalf, "The Minuteman Vulnerability Myth and the MX," *Strategic Review,* vol. 11, no. 2, Spring 1983, pp.7-8.

45. Trofimenko makes this point clearly in *Changing Attitudes Toward Deterrence:* "But if the mission of the military is to fight successfully and to win wars, then the mission of contemporary politicians is to prevent a nuclear war that can result in disaster for mankind" (p. 23). The Soviet view is portrayed sympathetically in Erickson, "The Soviet View of Deterrence." Erickson argues that the Soviet Union noted the U.S. theory of stability through MAD capabilities but was impressed by the reality of U.S. strategic force programs that exceed the requirements of MAD (p. 246). An important, penetrating recent analysis of the Soviet perspective is Jonathan S. Lockwood, *The Soviet View of U.S. Strategic Doctrine: Implications for Decision Making* (New Brunswick, N.J.: Transaction Books, 1983).

46. In the contexts of NATO policy over deployment of *Pershing* II and U.S. debate over the wisdom of deploying MX missiles in silos, Soviet spokesmen have been eloquent on the consequences of these developments for their missile-firing doctrine. Playing to Western anxieties, Soviet officials have suggested that they will be compelled to adopt some variant of a launch-on-warning or launch-under-attack firing rule.

47. Buchan, "The Anti-MAD Mythology," p. 14.

48. On this subject, see Henry S. Rowen, "The Evolution of Strategic Nuclear Doctrine," esp. p. 145; and "Formulating Strategic Doctrine," in Com-

mission on the Organization of the Government for the Conduct of Foreign Policy, vol. 4, appendix K: "Adequacy of Current Organization: Defense and Arms Control" (Washington: GPO, June 1975), pp. 219-34.

49. See Gray and Payne, "Victory Is Possible," pp. 17-18.

50. Soviet leaders do not believe that Soviet society should be a "hostage"—as envisaged in "the metaphysics of deterrence" as elaborated by Western scholars. See Erickson, "The Soviet View of Deterrence"; and Lockwood, *The Soviet View of U.S. Strategic Doctrine,* esp. chap. 10.

51. This was the basic thrust of James Schlesinger's argument in the period 1973-75. A useful period-piece, sympathetic to Schlesinger, was Richard Rosecrance, *Strategic Deterrence Reconsidered,* Adelphi Paper no. 116 (London: IISS, Spring 1975).

52. James Schlesinger was criticized in 1974-75 on virtually every ground save the correct strategic one, that LNOs would be very unlikely to "work." Critics waxed indignant over his alleged underestimation of likely U.S. casualties in a severely constrained counterforce exchange, but they tended overwhelmingly to neglect the point that U.S. LNOs would be unlikely to succeed in their policy purpose unless they were supported by forces expressing a "backstop" theory of escalation dominance.

53. John Erickson has noted that U.S. lack of interest in active and passive defense, far from reassuring the Soviet Union, has evoked suspicions that the United States is so confident of offensive counterforce success that it need not allocate resources for homeland defense ("The Soviet View of Deterrence," p. 246).

54. Erickson, "The Soviet Military System: Doctrine, Technology, and 'Style,' " p. 28.

55. LNOs (or for more recent terminology, SNOs [sub-SIOP nuclear options], could of course be executed against very discrete target sets in Eastern Europe.

56. On the selectivity debate over strategic targeting in the 1970s, see Lynn Etheridge Davis, *Limited Nuclear Options: Deterrence and the New American Doctrine,* Adelphi Paper no. 121 (London: IISS, Winter 1975-76); Warner R. Schilling, "U.S. Strategic Nuclear Concepts in the 1970s: The Search for Sufficiently Equivalent Countervailing Parity," *International Security,* vol. 6, no. 2, Fall 1981, pp. 48-79; and Freedman, *The Evolution of Nuclear Strategy,* chap. 25.

57. See Thomas Schelling, *Arms and Influence* (New Haven: Yale University Press, 1966), chap. 3. For an outstanding critical analysis of how wars can be lost if the United States thinks of military action not as *war* but rather as a process of diplomatic signaling and as a competition in risk-taking, see Stephen Peter Rosen, "Vietnam and the American Theory of Limited War," *International Security,* vol. 7, no. 2, Fall 1982, pp. 83-113.

58. See Jerome H. Kahan, *Security in the Nuclear Age: Developing U.S. Strategic Arms Policy* (Washington: Brookings Institution, 1975), esp. pp. 223-37.

59. On the strategic potential of SLBMs, see Joel S. Wit, "American SLBM: Counterforce Options and Strategic Implications," *Survival,* vol. 24, no. 4, July/August 1982, pp. 163-74.

60. See John Erickson, "The Soviet View of Nuclear War," Transcript of talk on BBC Radio 3, June 19, 1980, p. 10. For a useful and persuasive discussion of the importance the Soviet General Staff accords flexibility in command (in weapon employment) and flexibility in planning, see Joseph D. Douglass, Jr., and Amoretta Hoeber, *Conventional War and Escalation: The Soviet View* (New York: Crane, Russak [for the National Strategy Information Center], 1981), pp. 58-63. Also valuable is Lockwood, *The Soviet View of U.S. Strategic Doctrine,* esp. chap. 8.

61. There is no debate over the proposition that both Soviet and Western leaders view nuclear weapons as instruments of political intimidation, and that it is a rational act of policy so to employ these weapons. However, there is considerable disagreement over the extent to which one can describe Soviet military thoughts as "Clausewitzian." Do Soviet leaders regard war, even nuclear war, as "simply a continuation of political intercourse, with the addition of other means?" (Karl von Clausewitz, *On War,* ed. and trans. Michael Howard and Peter Paret [Princeton: Princeton University Press, 1976], p. 605). See also Erickson, "The Soviet View of Deterrence," p. 243; and P. H. Vigor, *The Soviet View of War, Peace, and Neutrality* (London: Routledge & Kegan Paul, 1975), esp. pp. 80-96.

62. See Jeffrey T. Richelson, "The Dilemmas of Counterpower Targeting," *Comparative Strategy,* vol. 2, no. 3, 1980, pp. 223-37; and Colin S. Gray, "Targeting Problems for Central War," *Naval War College Review,* vol. 33, no. 1, January-February 1980, pp. 3-21.

63. See Colin S. Gray, *Strategic Studies and Public Policy: The American "Experience"* (Lexington: University Press of Kentucky, 1982), chap. 4.

64. See Desmond Ball, "Counterforce Targeting: How New? How Viable?" *Arms Control Today,* vol. 11, no. 2, February 1981, pp. 1-2, 6-9.

65. See Colin S. Gray, "Presidential Directive 59: Flawed but Useful," *Parameters,* vol. 9, no. 1, March 1981, pp. 29-37.

66. Shocked, fearful, and possibly sick Soviet citizens are more likely to be desperate for the reassuring presence of familiar authority than they are to be angry to the point of revolt.

67. Note the comments in Harold Brown, *Department of Defense Annual Report, Fiscal Year 1982,* January 19, 1981, pp. 40, 42.

68. This theme was developed in Paul H. Nitze, "Deterring Our Deterrent," *Foreign Policy,* no. 25, Winter 1976-77, pp. 195-210.

69. Erickson, "The Soviet View of Deterrence," p. 247.

70. Scowcroft Commission Report, p. 6.

71. See William F. Scott and Harriet Fast Scott, *The Soviet Control Structure,* Final Report SPC 575 (Arlington, Va.: System Planning Corporation, April 1980). Paul Bracken is wide of the mark when he writes that: "What is especially worth noting about PD-59 and NSDD-13 is the idea advanced as the key to victory over the Soviet Union in a nuclear war, which is to target the political and military control system of the Soviet Union, not just its military forces. Soviet military and leadership targets had always been included in earlier plans. PD-59 for the first time declared open threats against them. As a consequence, a spate of academic articles appeared about the benefits of 'knocking out the Soviet control system.' Most of these were superficial, almost nonsensical, for they never even defined 'control system'; nor did they address what would happen if the Soviets shot back." (*The Command and Control of Nuclear Forces,* pp. 88, 89.) Speaking from personal experience, I can say that those of us who speculated in public about threatening the Soviet control system did have some very specific Soviet target sets in mind. Whether or not we should have defined those target sets in detail in the open literature is another matter. Furthermore, the U.S. official defense community has assumed for many years that Soviet targeteers assign a very high priority to attacking U.S. control targets. This author, among many of them, has argued repeatedly that the vulnerability of U.S. C^3I systems constitutes a potentially fatal Achilles heel.

72. The Central Committee *nomenklatura* are the holders of positions at the direct discretion of the Central Committee of the Soviet Communist party: in short, a system of highly centralized patronage.

73. There is considerable merit in the following argument advanced by Samuel P. Huntington: "The argument was sometimes made that 'killing the top Soviet leadership would leave the U.S. no one to fight the war with.' The

destruction of the central political leadership of the Soviet Union and of the command and communications channels by which Moscow exercises control over its military forces, and its satellites could well help precipitate the breakdown of Soviet authority that should be the wartime goal of American strategy." ("The Renewal of Strategy," in Huntington, ed., *The Strategic Imperative: New Policies for American Security* (Cambridge, Mass.: Ballinger Publishing Co., 1982), p. 33. Paul Bracken provides a contrasting perspective in *The Command and Control of Nuclear Forces,* esp. pp. 91-97. For example, Bracken alleges that "since actually destroying the Soviet leadership group would serve no military useful purpose, targeting is a bargaining tool that is intended to *influence* the group." (Emphasis in the original.) (p. 91.) It is by no means obvious that actually killing Soviet leaders would not serve a militarily useful purpose.

74. See Brown, *Department of Defense Annual Report, Fiscal Year 1982,* pp. 38-40; and Caspar W. Weinberger, Secretary of Defense, *Annual Report to the Congress, Fiscal Year 1984,* February 1, 1983, pp. 51, 57.

75. Weinberger, *Annual Report to the Congress, Fiscal Year 1984,* p. 55.

76. Save for very occasional lapses of self-discipline, administration spokesmen have sought to avoid conceding publicly their belief that the strategic nuclear balance of the early 1980s is in a condition of subparity to the U.S. disfavor. For example, Secretary Weinberger, in ibid., tells the truth, but not the whole truth, when he says, "We are now faced with a Soviet Union that has deprived us of our advantage in nuclear arms" (p. 20). A more accurate gauge of official belief is the argument in opposition to the proposal that the superpowers should first freeze, then reduce, their nuclear arms. Officials have asserted again and again that "freezing now" would preclude necessary force modernization. This author is fully cognizant of the many difficulties pertaining to assessment of the state of the strategic balance.

77. For example, see U.S. Army, Field Manual 100-5, *Operations* (Washington: Department of the Army, July 1, 1976).

78. As one would expect, these thoughts pervade the report of the Scowcroft Commission, which Harold Brown served assiduously as a consultant. See Scowcroft Commission Report, pp. 2, 6, 7.

79. In his last annual report, Harold Brown emphatically registered the point that Soviet economic targets have not been removed from the U.S. SIOP (*Department of Defense Annual Report, Fiscal Year 1982,* pp. 42, 43). See, also, Benjamin S. Lambeth and Kevin N. Lewis, "Economic Targeting in Nuclear War: U.S. and Soviet Approaches," *Orbis,* vol. 27, no. 1, Spring 1983, pp. 127-49; and Ball, *Targeting for Strategic Deterrence,* pp. 29-31.

80. See Holloway, *The Soviet Union and the Arms Race*, pp. 70-72. No less an authority than Sun Tzu advised, "What is of supreme importance in war is to attack the enemy's strategy" (*The Art of War*, trans. Samuel B. Griffith [Oxford: Oxford University Press, 1963], p. 77).

81. For negative commentaries on this point which at least understand the argument (even though they mischaracterize the world this author envisages), see Robert C. Gray, "The Reagan Nuclear Strategy," *Arms Control Today*, vol. 1, no. 2, March 1983, pp. 1-39; and Stanley R. Sloan and Robert C. Gray, *Nuclear Strategy and Arms Control: Challenges for U.S. Policy*, Headline Series no. 261 (New York: Foreign Policy Association, 1982), pp. 40-42.

82. A useful survey of the strategic doctrinal distance covered by the Carter administration (prior to PD-59) is Desmond Ball, *Developments in U.S. Strategic Nuclear Policy Under the Carter Administration*, ACIS Working Paper no. 21 (Los Angeles: Center for International and Strategic Affairs, UCLA, February 1980).

83. See Burt, "Reassessing the Strategic Balance," p. 50.

84. However, there is an important case to be made for defense mobilization planning as an integral part of U.S. (and NATO-Allied) defense policy. See Paul Bracken, "Mobilization in the Nuclear Age," *International Security*, vol. 3, no. 3, Winter 1978-79, pp. 74-93; Richard B. Foster and Francis P. Hoeber, "Limited Mobilization: A Strategy for Preparedness and Deterrence," *Orbis*, vol. 24, no. 3, Fall 1980, pp. 439-57; and Colin S. Gray, "Mobilization for High-Level Conflict: Policy Issues," in Robert L. Pfaltzgraff, Jr., and Uri Ra'anan, eds., *The U.S. Defense Mobilization Infrastructure: Problems and Priorities* (Hamden, Conn.: Shoe String Press, 1983), chap. 2.

85. See "President's Speech on Military Spending and a New Defense," quoted in *New York Times*, March 24, 1983, p. 20. For a useful compendium of newspaper reports concerning this speech and related matters, see "Star Wars," *Current News: Special Edition*, nos. 997 and 998 (May 4 and 5, 1983).

86. Schelling, *Arms and Influence*, pp. 69-91.

87. This belief pervaded Bernard Brodie's contribution to *The Absolute Weapon: Atomic Power and World Order* (New York: Harcourt, Brace, 1947), pp. 21-110.

88. On the fall of damage limitation from official favor, see Kahan, *Security in the Nuclear Age*, chap. 2; Fred Kaplan, *The Wizards of Armageddon* (New

York: Simon & Schuster, 1983), chap. 23; and Ball, *Targeting for Strategic Deterrence,* pp. 10-15.

89. It is at best a half-truth to assert, as Glenn Buchan does, "Once the Soviet Union decided to develop a credible strategic nuclear force, MAD was the inevitable result. There was nothing else that the United States could have done, short of taking the extreme course of risking an early nuclear war to stop the Soviet build-up" ("The Anti-MAD Mythology," pp. 14-15).

90. Quoted in Michael Howard, "On Fighting a Nuclear War," *International Security,* vol. 5, no. 4, Spring 1981, p. 14.

91. See Schilling, "U.S. Strategic Nuclear Concepts in the 1970s," pp. 64-65.

92. See Robert L. Arnett, "Soviet Attitudes Towards Nuclear War: Do They Really Think They Can Win?" *Journal of Strategic Studies,* vol. 2, no. 2, September 1979, pp. 172-91. But see Robert B. Berman and John C. Baker, *Soviet Strategic Forces: Requirements and Responses* (Washington: Brookings Institution, 1982), pp. 32-37; and Vigor, *Soviet Blitzkrieg Theory,* chap. 5. It is worth observing that even the "ground zero" organization handles this issue cautiously. See Ground Zero, *What About the Russians—And Nuclear War?* (New York: Pocket Books, 1983), pp. 169-70.

93. I have been criticized by Michael Howard for proceeding from a "denial of victory" to a "theory of U.S. victory" focus (Howard, "On Fighting a Nuclear War," p. 10).

94. U.S. Congress, Office of Technology Assessment, *The Effects of Nuclear War* (Washington: GPO, 1980), p. 3.

95. U.S. strategic forces have a major duty in extending deterrence. See Edward N. Luttwak, "The Problems of Extending Deterrence," in *The Future of Strategic Deterrence, Part I,* Adelphi Paper no. 160 (London: IISS, Autumn 1980), pp. 31-37; Earl C. Ravenal, "Counterforce and Alliance: The Ultimate Connection," *International Security,* vol. 6, no. 4, Spring 1982, pp. 26-43; and Anthony H. Cordesman, *Deterrence in the 1980s: Part I: American Strategic Forces and Extended Deterrence,* Adelphi Paper no. 175 (London: IISS, Summer 1982).

96. A very strong statement of this thesis is Robert S. McNamara, "The Military Role of Nuclear Weapons: Perceptions and Misperceptions," *Foreign Affairs,* vol. 62, no. 1, Fall 1983, pp. 59-80.

97. Relevant analyses on these and closely related subjects are Michael Howard, "Reassurance and Deterrence: Western Defense in the 1980s," *Foreign Affairs,* vol. 61, no. 2, Winter 1982-83, pp. 309-24; Eliot A. Cohen, "The

Long-term Crisis of the Alliance," *Foreign Affairs,* vol. 61, no. 2, Winter 1982-83, pp. 325-43; and Hedley Bull, "European Self-Reliance and the Reform of NATO," *Foreign Affairs,* vol. 61, no. 4, Spring 1983, pp. 874-92.

98. See Jack L. Snyder, *The Soviet Strategic Culture: Implications for Limited Nuclear Operations,* R-2154-AF (Santa Monica, Calif.: RAND, September 1977), pp. 39-40.

99. See Donald G. Brennan, "Soviet-American Communication Crises," *Arms Control and National Security,* vol. 1, 1969, pp. 81-88; and Ball, *Can Nuclear War Be Controlled?*

100. John Newhouse, *Cold Dawn: The Story of SALT* (New York: Holt, Rinehart & Winston, 1973), chap. 1.

101. See Gray, "National Style in Strategy"; and for a much broader discussion, Robert Dallek, *The American Style of Foreign Policy: Cultural Politics and Foreign Affairs* (New York: Alfred A. Knopf, 1983).

102. An exception to this rule has been registered over the "build-down" proposal for START. See William S. Cohen, "The Arms Build-Down Proposal: How We Got from There to Here," *Washington Post,* October 9, 1983, p. C-8.

103. The evidence is not yet in, but the price of victory for the Reagan administration on MX may be high. The prospects are not all gloomy. For a reasonably optimistic assessment, see Colin S. Gray, "Abiding Realities and Strategic Needs," *Air Force Magazine,* vol. 66, no. 7, July 1983, pp. 73-76.

104. A valuable and rare statement of a strategic case for defense is Benson D. Adams, "In Defense of the Homeland," *U.S. Naval Institute Proceedings,* vol. 109, no. 6, June 1983, pp. 44-49.

105. Donald G. Brennan, "The Case for Population Defense," in Johan J. Holst and William Schneider, Jr., eds., *Why ABM: Policy Issues in the Missile Defense Controversy* (New York: Pergamon Press, 1969), p. 116.

106. For an exceptionally strong affirmation of this case, see Adams, "In Defense of the Homeland." Adams believes that "strategic defenses were rejected not on strategic grounds but because they did not conform to the theories of mutual deterrence and arms control which themselves have proved invalid" (p. 49).

107. Little by way of substantial argument was advanced in favor of a defense dominant strategic environment during the 1970s. Following Donald Brennan's advocacy in the late 1960s (see Brennan, "The Case for Population Defense," esp. pp. 107-16), the public record was thin for many years. See Malcolm Wallop, "Opportunities and Imperatives of Ballistic Missile Defense," *Strategic Review,* vol. 6, no. 4, Fall 1979, pp. 13-21; and Ellory B. Block, "BMD's Role in National Survival and Recovery: A First Assessment," in Jake Garn et al., *The Future of U.S. Land-based Strategic Forces,* Special Report (Cambridge, Mass.: Institute for Foreign Policy Analysis, December 1980), pp. 64-80. However, true evidence of changing times was the appearance of Daniel O'Graham, *High Frontier: A New National Strategy* (Washington: High Frontier, 1982).

108. There is near unanimity among technical experts that this is the earliest time frame in which a strategic defense system that relied heavily on directed energy weapons could be deployed. I have discussed this issue at length in *American Military Space Policy: Information Systems, Weapon Systems, and Arms Control* (Cambridge, Mass.: Abt Books, 1983).

109. See Stanley Sienkiewicz, "Observations on the Impact of Uncertainty in Strategic Analysis," *World Politics,* vol. 32, no. 1, October 1979, pp. 90-110; and Benjamin S. Lambeth, "Uncertainties for the Soviet War Planner," *International Security,* vol. 7, no. 3, Winter 1982-83, pp. 139-66.

110. Adams, "In Defense of the Homeland," p. 48. Plainly Adams is contrasting defense and deterrence.

111. See Glenn Snyder, *Deterrence and Defense: Toward a Theory of National Security* (Princeton: Princeton University Press, 1961), pp. 3-15; and "Deterrence by Denial and Punishment," in Davis Bobrow, ed., *Components of Defense Policy* (Chicago: Rand McNally & Co., 1965), pp. 209-37.

112. Critics of U.S. BMD are correct in assuming that it is a two-power-or-none phenomenon. See Albert Carnesale, "Reviving the ABM Debate," *Arms Control Today,* vol. 11, no. 4, April 1981, p. 8.

113. The prospects for a nonnuclear defense of NATO Europe have been debated for more than twenty years. A useful review of ideas presently fashionable is Richard K. Betts, "Conventional Strategy: New Critics, Old Choices," *International Security,* vol. 7, no. 4, Spring 1983, pp. 140-62. See the correspondence between Edward N. Luttwak and Richard K. Betts triggered by this article in *International Security,* vol. 8, no. 2, Fall 1983, pp. 176-82.

114. A useful study of NATO's nuclear strategy is J. Michael Legge, *Theater Nuclear Weapons and the NATO Strategy of Flexible Response,* R-2964-FF (Santa Monica, Calif.: Rand Corporation, April 1983).

115. Schell, *The Fate of the Earth*.

116. E.g., Elizabeth Pond, "European Reflections on Reagan's 'Star Wars' Defense," *Christian Science Monitor,* April 11, 1983, p. 5.

117. These critical observations were gleaned by the author firsthand from conversations held in Europe following the president's speech of March 23, 1983, as well as from European news media.

118. See Thomas Karas, *The New High Ground: Systems and Weapons of Space Age War* (New York: Simon & Schuster, 1983), chaps. 6-8; and Gray, *American Military Space Policy*.

119. See Donald G. Brennan, "The Case for Missile Defense," *Foreign Affairs,* vol. 43, no. 3, April 1969, pp. 633-48.

120. The study panels established (following a presidential order of March 25, 1983) to explore the technology and policy dimensions of ballistic missile defense have recommended a range of funding alternatives for FY 1985-89 from $18 to $27 billion. See Clarence A. Robinson, Jr., "Panel Urges Defense Technology Advances," *Aviation Week and Space Technology,* vol. 119, no. 16, October 17, 1983, pp. 16-18.

Chapter 5

NUCLEAR STRATEGY: A NEW CONSENSUS?

Defense policy begins at home. Nuclear policy—strategy, programs, arms control positions—must first "work" in the minds of Americans if it is ever to be given the opportunity to "work" in the minds of Soviet leaders. In recognition of these self-evident if neglected truths, the Scowcroft Commission crafted its compromise package of interconnected proposals with a view to appealing to a wide domestic constituency. Consensus was the order of the day. "Finally, the Commission is particularly mindful of the importance of achieving a greater degree of national consensus with respect to our strategic deployments and arms control.[1]

For a nation that has no long-standing tradition in the field of military doctrine and that typically has tackled its defense problems pragmatically in an engineering spirit, the United States over the past fifteen years has been rent by the effects of an excess of zeal on the part of many devotees of rival strategic ideologies (or dogmas). The report of the Scowcroft Commission pointed to the following facts of recent history:

> For the last decade, each successive Administration has made proposals for arms control of strategic offensive systems that have become embroiled in political controversy between the Executive branch and Congress and between political parties. None has produced a ratified treaty covering such systems or a politically sustainable strategic modernization program for the U.S. ICBM force.[2]

Truth, in some abstract and scientifically testable sense, is not and cannot be the overriding goal of strategic inquiry. Defense analysts and theorists must recognize that the "best" is in practice all too often the enemy of the "good enough." Truth in a search for a sound U.S. nuclear policy is both political and strategic in character. The best the Scowcroft commissioners could accomplish was a set of proposals and arguments, each more or less interdependent, that had the overall character of the total enjoying

109

a measure of persuasiveness greatly superior to the sum of the separate parts. Good strategy is strategy that works, and in the first instance it must work at home.

The most basic reason the United States suffers (and benefits) from disruptive debates over fundamental issues of nuclear strategy is that there is no true authority for the guidance of policy. The most effective kind of authority is provided by the test of battle. The testing of armed forces in actual combat has a way of settling peacetime disputes among theorists. Fortunately for society, though unfortunately from the perspective of the quest for strategic truth, defense policies that succeed in preventing wars from occurring cannot be shown to have succeeded. It cannot be demonstrated beyond all argument, for example, that the substantial defense preparations of the United States and NATO Europe since 1950 have either prevented war or reduced the risk of war.[3] This author is not at all sympathetic to the opposing argument, that Western defense preparations (which leapt dramatically in scale following the invasion of South Korea) had a positive feedback effect on East-West political relations and substantially helped generate the perceived risk of war that is the basic rationale for the heavy defense preparations. But one cannot prove cause and effect between defense policies and wars deterred. No one knows what wars, if any, have been deterred since 1950.

One purpose of this study is to show the fragility of all arguments with respect to choices among nuclear strategies. Given the absence of the true authority of operational experience (i.e., nuclear war), the currency of debate is imagination, logic, and plausible argument. This author had no hesitation in identifying strategy Option Four (damage limitation for deterrence and coercion) as his clear first preference. In addition, he believes that Option Five (damage limitation with defense dominance) has great promise, while Option Three (counterforce and countercontrol preeminence) is seen as a way station that plainly is on the the right road. Those judgments notwithstanding, this study has sought to offer the most telling brief critical commentaries on all the strategy choices without prejudice or favor.[4] What emerges clearly from this exercise is that it is simpler to cast reasonable doubts on the prudence, utility, or technical feasibility of a strategy than it is to advance anything even approximating a wholly convincing case. If one wishes to appear wise in the halls of nuclear strategy and to secure a reputation as a truly responsible voice, one would be advised to advance no positive ideas of one's own on nuclear strategy, but rather to confine oneself to pointing to the problems with the ideas of others.

While there is an enduring need for skeptical review of ideas in the field of nuclear strategy, there also is a most immediate inescapable need for nuclear strategy. Skepticism is a virtue in a scholar or in a policymaker. But skeptical or not of all evident choices available, a policymaker (unlike a scholar) has to choose. As often as he can, he will choose as little as he can as late as he can—in the interest of preserving maximum flexibility to cope with the unknown and the unknowable. The truth is that U.S. (and Soviet) policymakers have been set an inalienable task that cannot be done efficiently. They must design a nuclear strategy for deterrence in a context of wholly inadequate information. Moreover, items of information that are critically important as a basis for prudent policymaking are literally unobtainable; further study will not provide the answers. Unfortunately, perhaps, the superpower policymaker is not at liberty to eschew choosing among nuclear strategies, and the various candidate weapons to express and implement those strategies, on the grounds that he does not know enough to make intelligent decisions.

The practical consequence of the situation described above is that policymakers, and many of their critics, defend their strategy and program preferences in terms that conceal, or attempt to conceal, the genuine unknowns and uncertainties. After all, if one is asking Congress to fund a weapon program that eventually will cost tens of billions of dollars, one feels compelled to sound as if one were confident that the weapon is needed. Critics of such a program are in turn obliged to affirm the absence of necessity for for the particular weapon in terms no less positive or strident. As many people have noticed over the years, the growth industry of systems analysis—particularly of quantitative systems analysis—has as a major function the role of concealing genuine uncertainties.[5]

Systems analysis has a knack for generating the wrong answers to the wrong questions. The law of the instrument applies widely. In this case one analyzes quantitatively things that lend themselves to such analysis. While better tools of analysis should enable one to allocate resources more efficiently, only strategic judgment (which in key regards can appeal only to inherently "soft" evidence) can advise as to the purposes for which resources should be allocated efficiently. Referring to the large-scale entry of what Bernard Brodie called the "scientific strategists" into the Defense Department in the early 1960s,[6] Eliot Cohen has observed, "What was new in 1960 was not the use of numbers or equations to help solve military problems but rather the coronation of one social science—economics—as the rightful queen of war planning and strategy."[7]

The purpose of this seeming digression is not to add another voice to the chorus of generally well-aimed criticism to which quantitative defense analysis, or, generically, systems analysis, has been subjected over the past fifteen years. Indeed, this author is not at all hostile to quantitative defense analysis, only to its misapplication and its tendency to evade, substitute for, or masquerade as genuinely strategic analysis. The purpose here is rather to advance the proposition that the critical uncertainties which are the basis for much of the controversy over nuclear strategy are genuine uncertainties that do not lend themselves to resolution through analysis, no matter how honestly and competently it is conducted.

Plainly, the U.S. defense community has a serious structural problem with regard to forging and sustaining that national consensus of which the Scowcroft Commissioners wrote. To repeat, the problem is one of lack of authority. There are three kinds of authority, which in principle if not always in practice could serve to discipline debate.

First, as already observed, there is the authority of experience, which is totally lacking in the nuclear field. Second, there is the authority of analysis. Unfortunately the relevant domain of analysis is subordinate to questions that can be answered only by judgment.[8] The central issue in defense preparation, "How much is enough?"[9] simply does not lend itself to quantitative analysis. For example, how does one proceed to answer such questions as "What quality and quantity of nuclear threat would be necessary to dissuade Soviet leaders from escalating a theater conflict that has become a stalemate?" and "What character of policy of all kinds would provide the necessary degree of credibility for U.S. threats to be genuinely dissuasive? Questions of this kind do not lend themselves to methodological assault invoking numerical calculation. In the first instance at least, the relevant data is of a qualitative kind, involving political, historical, psychological, and strategic judgments.

Ahead of time—that is to say ahead of that one day or week in twenty or thirty years when a Soviet leader or leaders, possibly facing a crisis-of-empire, either are or are not deterred—there is no way the U.S. government can be certain that its deterrent strategy and posture are truly adequate. The limitations of careful analysis are obvious when one considers the range of possible relationships between credible strength of political commitment and actual military power. Napoleon may not have said that the moral is to the material as three is to one, but the idea that an ounce of will may be worth a pound of capability is not to be despised.[10] The uniqueness of the rival human players, the distinctiveness of the historical circumstances, and the likely distinctiveness of the contending definitions and assertions of overlapping national interests all serve to provide major elements of indeterminacy with respect to judgments over how much is enough.

The third kind of authority in and over debate is the authority that a state may bestow on a particular institution or officeholder. A state may assign an institution (a division of a general staff organization, or an academy of military scholarship) or an individual (a chief of staff whose role is chief strategist; Count Alfred von Schlieffen, for example)[11] the duty of developing "correct," or at least authoritative, official doctrine. For reasons that transcend the scope of this discussion, the United States is bereft of organizations or individuals who may be said to have authority of this kind. In the United States, strategy-making is a bureaucratically diffuse activity that cannot be led for long or effectively from any single center of political power. Policy guidance can be provided by the White House, but guidance divorced from the ability to implement is always at risk from the activity, or lack of activity, of lower-level organizations that have parochial interests.

The U.S. system of government, with its constitutional and informal checks and balances reflecting U.S. history and political culture, cannot accommodate a genuine center of authority on nuclear strategy which could impose selected doctrine by fiat; such a situation would be incompatible with the extant political process. Moreover, no matter how convenient it would be for the orderly development of plans and programs, so great is the essential indeterminacy of some key questions pertaining to strategy that the unification of a truly effective authority in this field could lead to an attenuation of debate that would have an adverse effect on the quality of policy.

Chaotic though the process of strategy-making is in the United States, the policymaking and policy-implementing system is made to work by skilled politicians and officials, and there are powerful forces operating for continuity and even consensus. A democracy naturally functions politically on controversy, just as the news media in a democracy finds news value in the reality or appearance of discord. The debate over nuclear strategy and arms control of recent years reflects a condition that is both worse and better than typically portrayed. In commenting on the policy consequences of political controversy over nuclear issues, the Scowcroft Commission was right to claim, "Such a performance [the domestic U.S. political turmoil over SALT and the MX], as a nation, has produced neither agreement among ourselves, restraint by the Soviets, nor lasting mutual limitations on strategic offensive weapons."[12]

However, the U.S. policy problem goes deeper than may be supposed as the basis of observing that the nation has had Democratic, and then Republican, weapon systems and arms control proposals, with each new administration conducting close to a zero-base review of its strategic-program and strategic arms control inheritance from its predecessor. Underlying a good deal of the public debate over theories of deterrence, weapon systems, and approaches to strategic arms control is a set of basic questions that tend not to appear in their pure form. In their pure form they are so fundamental that they challenge the legitimacy or value of whole classes of activity that policymakers know they have no choice other than to pursue, whatever their inner doubts about the merit of the entire enterprise. Leading examples of these questions are the following:

1. *Can the idea of nuclear strategy, let alone nuclear strategy in action, have integrity?* Have nuclear weapons changed only the *char-*

acter of future war, as Soviet military theorists believe, or have they changed the *nature* of war itself, as U.S. officials and theorists generally assert? The U.S. government certainly would not agree with Lawrence Freedman's argument that:

> the position we have reached is one where stability depends on something that is more the antithesis of strategy than its apotheosis—or threats that things will get out of hand, that we might irrationally, that possibly through inadvertence we could set in motion a process that in its development and conclusion would be beyond human control and comprehension.[13]

But Freedman does point to the fact that uncertainty plays a central intended role in the deterrent value of NATO and U.S. nuclear strategy. Many people, not excluding responsible policymakers, wonder whether the architecture of nuclear strategy that has been constructed is not built on the sand of an unrealistic expectation concerning the nature of nuclear war. On the one hand, U.S. officials and theorists say that the nature of war has been changed by nuclear weapons,[14] while on the other hand they talk of plans for contingent employment of these weapons in terms compatible with the view that they were just another weapon. For example, the secretary of defense claims that "we need to be able to use force responsibly and discriminately, in a manner appropriate to the nature of a nuclear attack."[15]

The question this begs is whether nuclear force can be applied responsibly and discriminately. Is it possible that nuclear force by its very nature transcends responsible and discriminate application for strategic purposes?

Lest there be any misunderstanding, this author believes that the United States has no responsible choice other than to plan for the discriminate employment of nuclear force on behalf of clear political goals and overall in accordance with a "theory of victory."[16] However, there has been both too much polemic against nuclear weapons per se (as if scientific "progress" could be reversed) and too little study of just what nuclear weaponry means for the theory of war.

2. *What would be the character and consequences of nuclear war?* This study has referred many times to the uncertainty that surrounds attempts to answer this question. The U.S. government

today has a highly plausible theory of effective deterrence—in its offensive dimensions. But that theory is not complemented and sustained by a convincing story with regard to the survival of U.S. society in the event deterrence either fails catastrophically, or fails and then is painfully restored following a nuclear campaign of some intensity. Both the U.S. government and deterrence theorists at large tend to give the impression that they are not sensitive to the near-certain consequences of the nuclear action to which they lend contingent approval. The U.S. defense community has not confronted the basic question "Is nuclear war survivable?" Moreover, since no one can guarantee just what scale of violence would be unleashed, just how careful Soviet targeteers would be in their choice of DGZs,[17] heights of burst and weapon yields, and what the weather would be like on the day or days in question, defense planners are not at liberty to choose what may be termed preferred nuclear attacks. In short, is the United States in its nuclear strategy offering contingent promises to unleash a kind of war that it probably could not survive socially, economically, and politically?

3. *Is the arms race more to be feared than the Soviet Union?* Organizers of the nuclear freeze movement have encouraged the idea that the true enemy of U.S. security in particular and of humanity in general is the system of competitive armament rather than Soviet imperialism. The Scowcroft Commission tackled this issue vigorously as well. "Nor can we struggle against nuclear war or the arms race in some abstract sense without keeping before us the Soviet Union's drive to expand its power, which is what makes those struggles so difficult. We should face both problems directly."[18]

Underlying the arguments over weapon systems and approaches to negotiation is widespread uncertainty about the basic character of the world's first nuclear arms competition. Even people who have no difficulty endorsing the need to compete vigorously with the Soviet Union tend to wonder whether there is a dynamic to the technological competition which lacks meaning, particularly political and strategic meaning, beyond itself. Most Americans appear willing for their nation to do whatever is needed to ensure effective and stable deterrence over the long term, but as the dramatic surge of grassroots support for the idea of a nuclear freeze "now" illustrated, they are troubled lest somehow the system of competitive armament has escaped true political management in support of prudent strategic goals. The difficulty here is that to date neither officials nor scholars have provided a convincing and easily explained theory of what mix of elements, in what interrelation, drive the arms competition. In their excellent citizen's guide *Living with Nuclear Weapons,* the Harvard Nuclear Study Group recognized the ambivalence and confusion on this subject and offered five theories, or classes of explanation, of what drives the competition. Understanding is hampered by the reality of apparently redundant causation. Nonetheless, the Harvard authors did succeed in isolating the most significant of driving factors. "So long as the political competition between the United States and the Soviet Union continues in its present mode, statesmen on both sides will continue to be driven to say, 'We have no choice. We'll go ahead.' "[19]

4. *Is the East-West arms control process incapable of having a net beneficial impact on national and international security?* It is a political fact of life in the United States of the early 1980s (and Western Europe) that while one may challenge the sense in particular arms control proposals, it is not legitimate or even respectable to challenge the merit of the arms control process as a whole. However, doubts about the value of the process do exist, and those doubts, and speculation concerning those doubts, have an influence over public debate. The U.S. and NATO-European publics believe that the arms control process is important. Absurd though it really is, evidence of "belief" in and commitment to arms control has come to be required for a reputation as a responsible leader. It is difficult for a government to approach the prospect of, let alone behave in, arms control negotiations in a prudent manner when its domestic (and allied) continuency demands evidence of commitment and perhaps evidence of "progress," while the adversary-partner holds to an instrumental view of the arms control process as a tool of political warfare.[20]

Many of the elements of strategic planning and nuclear strategy discussed in this study have an actual or potential arms control dimension. This has been demonstrated clearly in the report of the Scowcroft Commission. The commissioners stated, "It is a legitimate, ambitious, and realistic objective of arms control agreements to channel the modernization of strategic forces, over the long term, in a more stable direction than would be the case without such agreements."[21]

The commissioners had no practicable choice. They knew that their interlocking package of force modernization proposals would not be politically acceptable unless it was accompanied by a pervasive arms control story. The general public would have difficulty trying to glean from the words of politicians, officials, and defense commentators the fact that there are serious reasons to question the value for security of the East-West arms control process. These reasons include the effect of the sharply contrasting characters of the Soviet and U.S. systems of government; the dynamic for competition that flows from a deep-seated political rivalry (which is both geostrategic and ideological in nature); the unique difficulty a democracy has in negotiating competently with a state that has no popular domestic constituency to satisfy; the different duties with which the two sides charge their strategic forces; the different doctrines that guide postural development; and the potential of arms control regimes to misdirect postural evolution.[22]

Suffice it to say that there are grounds far removed from a visceral anti-Sovietism, or a conviction that the Soviets are not to be trusted, for questioning the net value of the arms control process. Senators who press an administration to provide evidence of commitment to or faith in arms control, as though this were a theological issue ("redemption through arms control") rather than a matter of statecraft, exemplify the reasons many thoughtful people question the basic ability of a Western democracy to conduct itself prudently in an arms control process.

5. *Are nuclear weapons morally acceptable?* Whatever merit, or otherwise, one finds in the heavily negotiated text of the pastoral letter on war and peace of the National Conference of Catholic Bishops,[23] the letter brought to public consciousness a set of issues that has long troubled so-called defense professionals. For example, ethical concern was an important aspect of the criticism by generally conservative strategic thinkers of the ideas of mutual assured vulnerability and mutual assured destruction. It is true that critics of MAD tended to employ moral argument more as an apparent makeweight to strategic claims than as a powerful

point on its own, but still many U.S. strategic thinkers have been discomforted by the idea that it was the United States and not the Soviet Union that upheld the strategic merit for deterrence in holding tens of millions of civilians at nuclear risk. The Catholic bishops wrote:

> These considerations [particular complexities and paradoxes] of concrete elements of nuclear deterrence policy, made in light of John Paul II's evaluation, but applying it through our own prudential judgments, lead us to a strictly conditioned moral acceptance of nuclear deterrence. We cannot consider it adequate as a long-term basis for peace.[24]

Thoughtful strategists tend to agree with this judgment, or at least with the second sentence. In common with most strategic thinkers, the Catholic bishops fear that one day the deterrence system will fail. President Reagan spoke to this fear when he urged on March 23, 1983, that the nation's defense scientists examine the prospects for ballistic missile defense.[25]

There is some danger that the pastoral letter, notwithstanding the great care with which it was drafted and redrafted, could have a net negative impact on peace and stability. It is possible that the conditional approval it reluctantly bestows on the nuclear deterrence system is so conditional and faint-hearted that it will undermine the legitimacy of the system in the minds of many people. For example, the bishops place heavy emphasis on the need for arms control. Although the actual language of the letter is hedged with many "ifs" and "buts," it is not difficult to see how the United States, despite sincere and energetic activity, might be held not to have tried hard enough. One thing at least is certain: The authors of the letter framed their arguments in ways that are pertinent to the making and execution of strategic policy. Even where a strategist may disagree with the bishops—as, for example, this author does with the insistence of the letter on a "no first use" stance[26]—it is gratifying to see that the letter does accurately recognize the nature and much of the detail of views different from its own.

Defense professionals both in and out of uniform should welcome the initiative of the pastoral letter. They should be sensitive to moral questions and recognize responsible and reasoned skeptical analysis as moral grounds for what it is—a necessary dimension to strategic thought and action.

In this final chapter much has been made of uncertainty, confusion, and regions of dispute. Those elements are indeed present in the realm of nuclear strategy and arms control, but it would not be accurate to claim that all is chaos. On the contrary, it is remarkable how much agreement there is on some key elements of strategy and strategic planning in the whole nuclear policy field, including arms control. It is the contention of this study that the nuclear strategy of the United States at the present time—at the level of doctrine at least—is in good and improving condition. That optimistic contention must be balanced by the fact that today and for years to come actual U.S. strategic capabilities will lag far behind the requirements that flow logically from doctrine.

There is a cast of characters on the theological arms control wing of the U.S. defense community who oppose virtually every new weapon idea under consideration for adoption. The consensus referred to here plainly does not and never will include such people as Paul Warnke, Herbert Scoville, Richard Garwin, and Jeremy Stone.[27] Moreover, it may be argued that the mainstream of a defense community benefits from the activities of critics on the left and right fringes of informed opinion.[28] Consensus is valuable, indeed essential, for moving the ship of state in a reasonable, orderly way. But the widespread agreement and shared assumptions do not mean that the agreements and assumptions are correct. At different times in the past there was a consensus on such matters as the flatness of the earth, the physical integrity of the atom, the brevity of the next general war (in 1914), and the impracticability of ballistic missiles as useful weapons of intercontinental warfare.

What has emerged over the past decade, and most dramatically over the past several years, is a working though far from unchallenged consensus within the U.S. defense community on the necessity for what may be termed "war-fighting deterrence."[29] This trend was evident in the evolution of policy guidance from President Nixon's National Security Decision Memorandum 242 of 1974, through President Carter's PD-59 of 1980, to President Reagan's National Security Decision Directive 13 of

1981. However, the extent of the shift of the center of gravity of opinion over nuclear strategy, as between threats to punish Soviet society and threats to defeat the Soviet state and its military instruments, became fully manifest only in 1983 in the context of the public debate over the report of the Scowcroft Commission.

Critics from the left and the right of the political spectrum passed over in near silence the nuclear strategy that the commissioners specified in blunt language. Senators and editorial and op-ed writers expressed skepticism over the merit of having the hard-target punch of the MX ICBM in a basing scheme that was not independently survivable. Many people speculated about how interested the commissioners really were in the small single-warhead ICBM, and many more expressed skepticism over the real willingness of President Reagan to behave with energy, expedition, and imagination in the field of strategic arms control. What is surprising is the debate that did not occur.

People have not noticed, perhaps do not care, or possibly approve of the fact that the commissioners described the hypothetical small ICBM as follows: "It should have sufficient accuracy and yield to put Soviet hardened military targets at risk."[30] Much earlier in the report, the commissioners wrote: "Deterrence is not, and cannot be, bluff. In order for deterrence to be effective we must not merely have weapons, we must be perceived to be able, and prepared, if necessary, to use them effectively against the key elements of Soviet power."[31]

It is true that the report identified stability as "the primary objective" of U.S. strategic-force modernization of arms control proposals.[32] But the stability envisaged by the commissioners is a stability that is a light-year distant from the stability theme dominant in the defense community at the time of the negotiation of SALT I in the period 1969-72.[33] As recently as 1978-79, the theory of deterrence outlined in the report of the Scowcroft Commission was widely viewed as radical, even radically dangerous. This author was subjected to a great deal of criticism for saying and writing in the late 1970s what today is apparently orthodox wisdom among both Republicans and Democrats.[34]

The working consensus evident from the absence of public debate over the commissioners' assertion that "deterrence . . . requires military effectiveness"[35] remains limited to the scope of

Option Three of chapter 4 above, "counterforce and countercontrol preeminence with recovery denial"—in other words, to a theory of victory denial in terms deemed consistent with Soviet "thoughtways."[36] There is as yet no consensus as to the merit of the United States' pursuing a damage-limiting strategy. As the public popularity of the idea of defense of the U.S. homeland suggests,[37] a working consensus within the defense community should be attainable and sustainable once there is a widespread belief in the technical feasibility of the task. Such belief is far from widespread today. To date, proponents of damage limitation through a mixture of offensive and active and passive defensive means have tended to be more eloquent at the level of philosophy than they have been persuasive at the level of proposed engineering application.

In the interest of civility in debate, not to mention the avoidance of misrepresentation, the time is long past for proponents of homeland defense to cease to fulminate against alleged arms control ideologues who favor a MAD theory of strategic stability. Of course there are closeminded people who are beyond rational argument on the issue of the requirements of stable deterrence, and such people exist on all sides of the nuclear strategy debate. It is essential to recognize, however, that the largest number of opponents and skeptics of damage-limiting and defense-dominant strategies are in opposition or are skeptical for good and certainly honest reasons. This author believes that their reasons are not good enough and that they should elevate their vision, but contrary to the impression one can derive early from reading the advocacy literature on nuclear strategy and arms control, the world is not divided neatly into strategic thinkers who wear white hats and those who wear black hats. A defense community that can shift over the course of a decade from a societal-punishment theory of deterrence—as the public doctrinal leitmotif, not as the reality of force allocation in target planning—to a thoroughgoing counterforce-emphasizing theory of deterrence should be a defense community fully capable of seeing the wisdom of balancing, indeed complementing, offensive threats to Soviet state power with a plausible multitiered theory of defense for the U.S. homeland.

Notwithstanding the uncertainties over some fundamental issues that exist beneath the surface of U.S. strategic nuclear planning and nuclear strategy, the elements of consensus are large and appear to be growing. The success of the Scowcroft Commission was something of a surprise to the rank and file of commentators because they were so habituated to evidence of dispute and were so hardened to policy failure with respect to the successive changing details of the MX ICBM programs that they did not comprehend the breadth of shared assumptions available for a truly politically skilled group to shape into a coherent story. Without wishing to diminish the accomplishments of the Scowcroft Commission or the political intelligence and ability of President Reagan, one must understand the material with which the commission had to work, from which a bipartisan consensus could be forged.

First, there is very general agreement on the goals of U.S. foreign policy and on the security commitments of different kinds that express and support that policy. The commission asserted that the United States must "contain" "the threat of aggressive totalitarianism" and then proceeded to demonstrate the relevance of the strategic-forces balance, and of ICBMs specifically, to that duty.[38] This author has seen no noteworthy criticism or even skepticism concerning the appropriateness of the foreign policy reasoning that frames the report.

Second, despite the doubts of many intellectuals and even some defense professionals, there is no debate of immediate policy-relevance over the necessity for nuclear strategy per se. Strategic planning and orderly weapon acquisition is not threatened by controversy over the fundamental sense in the enterprise. Most people are to a degree skeptical, or agnostic, over the likely match between prewar plans and operational reality, but that skepticism is not novel to the nuclear age or to nuclear planning in the nuclear age.

Third, the U.S. defense community and the U.S. political system more broadly are firmly and irrevocably committed to the doctrinal preeminence of the concept of deterrence. The debate over U.S. nuclear strategy—among society-punishers, with or without targeting flexibility, and counterforce advocates with or without a damage-limitation story for homeland defense—almost entirely fits under the umbrella of deterrence. Even those who worry about the failure of deterrence tend to frame their recommendations with a view to its restoration.

Fourth, the commission was able to invoke the somewhat elusive tribal totem of "stability" both as claimed source of inspiration and as major objective.[39] Stability means all things to all people, which is its principal virtue as a symbol for manipulation.[40] The Scowcroft Commission, whose master was political necessity and not a board of peers of strategic theorists, argues a stability case for offsetting Soviet ICBM firepower by basing the MX in silos. In addition, they argue that the small ICBM is likely both to diminish Soviet incentives to compete and to reduce Soviet incentives to attack first during an acute crisis. Overall, the commission provides a skillfully woven potpourri of stability arguments—political stability, technical weapon-system stability, crisis stability, and arms control stability—some of which are more than a little shaky if assayed in isolation. But as the commission intended and achieved, strategic-force and arms control issues would be considered as a whole.

Fifth, with greater or lesser enthusiasm, everybody seems to favor flexibility and control in strategic nuclear planning. The criticism to which Defense Secretary James Schlesinger was subject in 1974-75 with respect to the desirability of flexibility for waging limited central war[41] has not been repeated in recent years. Deep skepticism remains, indeed has probably grown, over the feasibility of maintaining control over nuclear forces in wartime, but it is not respectable today to oppose flexibility per se.

Sixth, counterforce and countercontrol targeting have their critics and skeptics, but—as shown by the absence of comment on this truly central aspect of the commission report—the U.S. defense community and its pertinent constituencies have accepted the denial-of-victory logic of Harold Brown's "countervailing strategy."[42] Even people who are seriously troubled by what they see as the instability potential of so-called war-fighting theories of deterrence do not as a general rule challenge the argument that effective deterrence must rest on posing threats that speak to Soviet anxieties. No one has offered a strong challenge to the proposition that the most essential assets of the Soviet state, in the eyes of Soviet leaders, are the political control structure of the state and its military and paramilitary instruments of external and domestic coercion. The countermilitary and countercontrol strategy of the final years of the Carter administration, and now of the

Reagan administration, flow directly from this theory of deterrence, as does the requirement for a weapon like the MX ICBM, which can hold those Soviet assets confidently and promptly at risk.

Seventh, there is a working consensus on the necessity for careful exploration of the technical feasibility of strategic defense. While there is some continuing doctrinal opposition to the very idea, most people seem on the one hand genuinely to be agnostic on the crucial issue of technical feasibility, and on the other hand to be unwilling to endorse strategic conceptions that depend on as-yet-undemonstrated technologies. Notwithstanding the temporally parallel drafting of the report of the Scowcroft Commission and the president's speech of March 23, 1983, it is worth noting that the report makes only conventional, perfunctory, and even discouraging reference to ballistic missile defenses,[43] and makes no reference to defense of the homeland or to damage limitation as a prospectively important complement to offensive nuclear capability. At the present time there is no consensus on the need for or feasibility of strategic defense—that is, of comprehensive defense of the U.S. homeland—but it is not controversial to assert a need to explore technological possibilities.

Eighth, there is a strong political consensus about the necessity for sustaining a visible East-West arms control process. At the level of practical politics, if not always of intellectual conviction, the conduct of arms control activity is not controversial. Motives and approaches to arms control differ widely. Some people see the need for the United States to believe in arms control so that it cannot be blamed should progress not be achieved. This is a political damage-limiting perspective. Other people believe that the arms control process can have the positive result of producing modest but useful agreements that enhance predictability, help channel weapons modernization away from areas that promote the highest of anxieties, possibly save scarce defense dollars, and generally have a positive supportive influence on East-West political relations.[44] Strategic logic and political judgment meet in a consensus that agreements cannot be secured which are to the net disadvantage of one side; that the strategic balance cannot be restructured, and major-scale force reductions are not within the bounds of possibility; and that neither side will agree to an arms control regime that would mean a significant change in either superpower's theory of war. Unfortunately there is no consensus among Western publics in support of the proposition that the Soviet Union has a unified theory of conflict and that it approaches an arms control process as an institution that has no value in and of itself but that has seemingly perennial utility as one among many instruments for the conduct of political warfare.

This study has deliberately steered away from discussion of the merits of particular weapon programs. In the context of considering some of the realities of strategic planning, the author has suggested that there is order beneath the apparent chaos of strategy and weapons debate. Furthermore, he has suggested that broad strategy choices are distinguishable and that despite some irreducible uncertainties on truly fundamental matters, the United States enjoys a consensus on assumptions and risks sufficient to design and conduct a nuclear strategy that has strategic merit and that meets the tests of domestic and allied political acceptability.

As the technology of strategic defense matures—as a consequence of selective but generous nurture—the United States should be prepared to effect an orderly transition to a strategic force posture that would be balanced between offensive and defensive capabilities. As to the present condition of U.S. strategic programs, the Scowcroft Commission has provided a flag around which people of differing perspectives may rally. But the possibility still exists that excessive enthusiasm in pursuit of particular vision of "the best," not to mention less savory political motives, may yet hinder the program action that requires cooperation between the different branches of government. By way of a closing judgment, it would be difficult to improve on the following words written by Anthony Cordesman: "This is, however, not the time for more division over ideals. There is a time for discussion and a time for action, but in the case of strategic forces, the time for discussion of options has run out. It is time for a national act of will."[45]

NOTES

1. *Report of the President's Commission on Strategic Forces,* April 1983, p. 25 (hereafter cited as Scowcroft Commission Report).

2. Ibid.

3. But see Morton H. Halperin, "Keeping Our Troops in Europe," *New York Times Magazine,* October 17, 1982, pp. 82-84, 86, 88, 93-97.

4. The author claims only virtuous intent. He recognizes that devotees of strategy preferences different from his own may believe that their cases were not fairly presented or criticized.

5. E.g., see James R. Schlesinger, *Selected Papers on National Security, 1964-68* (Santa Monica, Calif.: RAND, September 1974); and Gregory Palmer, *The McNamara Strategy and the Vietnam War: Program Budgeting in the Pentagon, 1960-1968* (Westport, Conn.: Greenwood Press, 1978).

6. Bernard Brodie, "The Scientific Strategists," in Robert Gilpin and Christopher Wright, eds., *Scientists and National Policy Making* (New York: Columbia University Press, 1964), pp. 240-56.

7. Eliot A. Cohen, "Guessing Game: A Reappraisal of Systems Analysis," in Samuel P. Huntington, ed., *The Strategic Imperative: New Policies for American Security* (Cambridge, Mass.: Ballinger Publishing Co., 1982), p. 166.

8. For example, in 1963-64 the preeminent question "Should we fight in Vietnam?" was quintessentially of a strategic character and could be answered only by judgment.

9. Claims to the contrary pervade Alain C. Enthoven and K. Wayne Smith, *How Much Is Enough? Shaping the Defense Program, 1961-1969* (New York: Harper & Row, 1971).

10. There is a great deal to be said for Herman Kahn's opinion that "usually the most convincing way to look willing is to be willing" *(On Thermonuclear War* [Princeton: Princeton University Press, 1960], p. 287).

11. Chief of the German general staff, 1891-1905.

12. Scowcroft Commission Report, p. 25.

13. Lawrence Freedman, *The Evolution of Nuclear Strategy* (London: Macmillan & Co., 1981), p. 400.

14. Scowcroft Commission Report, pp. 1, 2.

15. Caspar W. Weinberger, Secretary of Defense, *Annual Report to the Congress, Fiscal Year 1984,* February 1, 1983, p. 54.

16. Colin S. Gray, "Nuclear Strategy: The Case for a Theory of Victory," *International Security,* vol. 4, no. 1, Summer 1979, pp. 54-87.

17. DGZ: designated ground zero, a point on the ground which is, or is directly below, the aiming point.

18. Scowcroft Commission Report, p. 1.

19. Harvard Nuclear Study Group, *Living with Nuclear Weapons* (New York: Bantam Books, 1983), p. 113. The five theories cited by the Harvard authors are: genuine security requirements; uncertainty and misperceptions; domestic political pressures; technological determinism; and international political rivalry (pp. 111-13). For detailed treatments of the question of the dynamics of the nuclear arms race, see Colin S. Gray, *The Soviet-American Arms Race* (Lexington, Mass.: Lexington Books, 1976); and Miroslav Nincic, *The Arms Race: The Political Economy of Military Growth* (New York: Praeger Publishers, 1982).

20. See Colin S. Gray, *Arms Control: Problems,* Information Series no. 132 (Fairfax, Va.: National Institute for Public Policy, January 1983); and Keith B. Payne and Dan Strode, *Arms Control: The Soviet Approach and Its Implications,* Information Series no. 142 (Fairfax, Va.: National Institute for Public Policy, April 1983).

21. Scowcroft Commission Report, p. 22.

22. See Laurence Martin, *The Two Edged Sword: Armed Force in the Modern World* (London: Weidenfeld & Nicolson, 1982), pp. 72-73.

23. National Conference of Catholic Bishops, "The Challenge of Peace: God's Promise and Our Response," *Origins,* National Conference documentary service, vol. 13, no. 1, May 19, 1983. (Hereafter cited as Pastoral Letter.)

24. Ibid., p. 18.

25. "President's Speech on Military Spending and a New Defense," *New York Times,* March 24, 1983, p. 20.

26. Pastoral Letter, p. 15.

27. The integrity and knowledge of these people are not challenged here.

28. A community of defense professionals who share many assumptions in common have to be compelled by outside challenge from time to time to reconsider the continuing validity of some of those assumptions. A Herbert Scoville and Richard Garwin on the liberal side, and a William Van Cleave and Edward Luttwak on the conservative side, are both needed to throw the occasional rock into pools of complacency and intellectual laziness.

29. See Keith B. Payne, *Nuclear Deterrence in U.S.-Soviet Relations* (Boulder, Colo.: Westview Press, 1982). Pejorative and somewhat inaccurate description of this phenomenon is provided in Robert Scheer, *With Enough Shovels: Reagan, Bush, and Nuclear War* (New York: Random House, 1982). Rather better is Thomas Powers, "Choosing a Strategy for World War III," *The Atlantic,* November 1982, pp. 82-110.

30. Scowcroft Commission Report, p. 15.

31. Ibid., pp. 2-3.

32. Ibid., p. 3.

33. See John Newhouse, *Cold Dawn: The Story of SALT* (New York: Holt, Rinehart & Winston, 1973), chap. 1.

34. E.g., Bernard Brodie, "The Development of Nuclear Strategy," *International Security,* vol. 2, no. 4, Spring 1978, p. 83; and McGeorge Bundy, "Strategic Deterrence Thirty Years Later: What Has Changed?" in *The Future of Strategic Deterrence: Part 1,* Adelphi Paper no. 160 (London: IISS, Autumn 1980), p. 6.

35. Scowcroft Commission Report, p. 7.

36. Ken Booth, *Strategy and Ethnocentrism* (London: Croom, Helm, 1979), p. 14.

37. White House mail ran overwhelmingly in favor of the president's suggestion of homeland defense on March 23, 1983.

38. Scowcroft Commission Report, p. 1.

39. Ibid., p. 3.

40. See Colin S. Gray, "Strategic Stability Reconsidered," *Daedalus,* vol. 109, no. 4, Fall 1980, pp. 135-54.

41. See Herbert Scoville, "Flexible MADness," *Foreign Policy,* no. 14, Spring 1974, pp. 164-77; and George Rathjens, "Flexible Response Options," *Orbis,* vol. 18, no. 3, Fall 1974, pp. 677-88.

42. See the testimony of Harold Brown in U.S. Senate, Committee on Foreign Relations, *Nuclear War Strategy, Hearing,* 96th Cong., 2d sess. (Washington: GPO, September 16, 1980 [top secret hearing, sanitized, and released February 18, 1981]).

43. Scowcroft Commission Report, pp. 9-10.

44. The "modest merit" school is well represented in Leslie H. Gelb, "A Practical Way to Arms Control," *New York Times Magazine,* June 5, 1983, pp. 33-42; and Harvard Nuclear Study Group, *Living with Nuclear Weapons,* chap. 9.

45. Anthony H. Cordesman, "The President's Commission on Strategic Forces: The Need for a National Act of Will," *Armed Forces Journal International,* May 1983, p. 90.